MTH 086: Introductory Algebra

Aufmann | Barker | Lockwood

 CENGAGE

Australia • Brazil • Mexico • Singapore • United Kingdom • United States

MTH 086: Introductory Algebra

Mathematics: Journey from Basic Mathematics through Intermediate Algebra
Richard N. Aufmann | Joanne S. Lockwood

© 2016 Cengage Learning. All rights reserved.

For product information and technology assistance, contact us at
Cengage Learning Customer & Sales Support, 1-800-354-9706.

For permission to use material from this text or product, submit all requests online at **www.cengage.com/permissions.**
Further permissions questions can be emailed to
permissionrequest@cengage.com.

This book contains select works from existing Cengage learning resources and was produced by Cengage learning Custom Solutions for collegiate use. As such, those adopting and/or contributing to this work are responsible for editorial content accuracy, continuity and completeness.

Compilation © 2017 Cengage Learning

ISBN: 978-1-337-05299-3

Cengage Learning
20 Channel Street
Boston, MA 02210
USA

Cengage Learning is a leading provider of customized learning solutions with employees residing in nearly 40 different countries and sales in more than 125 countries around the world. Find your local representative at: **www.cengage.com.**

Cengage Learning products are represented in Canada by Nelson Education, Ltd.

For your course and learning solutions, visit **www.cengage.com.**

Purchase any of our products at your local college store or at our preferred online store **www.cengagebrain.com.**

Visit our custom book building website at **www.compose.cengage.com.**

Brief Contents

Module 1: Whole Numbers

Objective 1.1A: Identify the order relation between two numbers

Natural Numbers: $1, 2, 3, 4, \ldots$

Whole Numbers: $0, 1, 2, 3, 4, \ldots$

The **graph** of a whole number is shown by placing a heavy dot on a **number line** directly above the number.

Examples:

1. Graph 7 on the number line.

   Solution:

   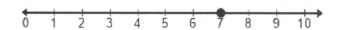

2. On the number line, what number is 5 units to the left of 9?

   Solution:

   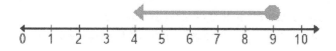

   4 is 5 units to the left of 9.

On a number line, the numbers get larger as we move from left to right. A number that appears to the right of a given number is **greater than** the given number (symbol: >). A number that appears to the left of a given number is **less than** the given number (symbol: <).

An **inequality** expresses the order of two mathematical expressions. $5 < 9$ and $8 > 7$ are inequalities.

Examples:

1. Place the correct symbol, < or >, between 341 and 89.

   Solution: $341 > 89$

2. Place the correct symbol, < or >, between 0 and 179.

   Solution: $0 < 179$

3. Arrange the numbers 3, 3,330, 303, 3003, and 330 in order from smallest to largest.

   Solution: 3, 303, 330, 3003, 3,330

Practice Exercises 1.1A:

1. Graph 2 on the number line.

   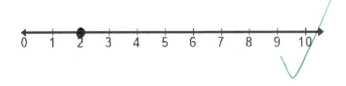

2. On the number line, what number is 6 units to the right of 2?

   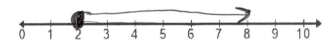

3. Place the correct symbol, < or >, between 1,049 and 109.

   1,049 > 109

4. Place the correct symbol, < or >, between 4,610 and 4,061.

   4,610 > 4,061

5. Place the correct symbol, < or >, between 918 and 8,978

   918 < 8,978

6. Arrange the numbers 1,331, 313, 3,311, 3,113, and 1,111 in order from smallest to largest.

313, 1,111, 1,331, 3113, 3311

7. Arrange the numbers 72, 48, 84, 93, and 13 in order from smallest to largest.

13, 48, 72, 84, 93

Objective 1.1B: Write whole numbers in words, in standard form, and in expanded form

Standard Form: When a whole number is written using the digits 0 through 9 it is said to be in **standard form**. The position of each digit in the number determines its place value.

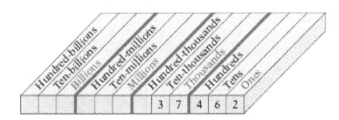

Examples:

1. In the number 37,462, what is the place-value of the digit 4?

   Solution: The place-value of the digit 4 is hundreds.

2. In the number 7,540,312,269, what is the place-value of the digit 4?

   Solution: The place-value of the digit 4 is ten-millions.

Each group of digits separated by a comma is called a period. To write a number in words, start from the left. Name the number in each period. Then write the period name in place of the comma.

Example:

Write 4,301,243,890 in words.

Solution: four billion three-hundred one million, two hundred forty three thousand, eight hundred ninety

To write a whole number in standard form, write the number named in each period, and replace each period name with a comma.

Example:

Write seventy-two billion, two million, three hundred nine thousand, twelve in standard form.

Solution: 72,002,309,012

A whole number may also be written in **expanded form**.

Example:

Write 48,308 in expanded form.

Solution: 40,000 + 8,000 + 300 + 8

Practice Exercises 1.1B:

1. In the number 2,401,879, what is the place value of the digit 0?

   ten thousands

2. Write 3,049,000,246 in words.

   three billon, forty nine thousand two hundred forty six

3. Write 12,345,678,900 in words.

   twelve billon three hundred forty five, six hundred seventy eight thousand, nine hundred

4. Write seven million eight in standard form.

   7,000,008

5. Write eighty-two billion, four hundred eighty million, three hundred fifteen thousand, seventy-two in standard form.

   82,480,315,072

6. Write 58,030,241 in expanded form.

   50,000,000 + 80,000,000 +   + 30,000 + 200 + 400 + 1

7. Write 7,245 in expanded form.

   7000 + 200 + 40 + 5

Objective 1.1C: Round a whole number to a given place value

Rounding: Giving an approximate value for an exact number is called rounding. We can round a number to a given place value without using the number line by looking at the first digit to the right of the given place value.

Examples:

1. Round 8,271 to the nearest hundred.

   Solution: In 8,271, the first digit to the right of the 2 in the hundreds place is 7. Since the digit to the right of the given place value is greater than or equal to 5, increase the digit in the given place value by 1, and replace all other digits to the right by zeros.

   8,271 rounded to the nearest hundred is 8,300.

2. Round 4,239,876 to the nearest hundred-thousand.

   Solution: In 4,239,876, the first digit to the right of the 2 in the hundred-thousands place is 3. Since the digit to the right of the given place value is less than 5, replace that digit and all digits to the right of it by zeros.

   4,239,876 rounded to the nearest hundred-thousand is 4,200,000.

3. Round 4,895 to the nearest ten.

   Solution: In 4,895, the first digit (the only digit!) to the right of the 9 is 5. The digit to the right of the given place value is greater than or equal to 5, so increase the digit in the given place value by 1. Since 9 + 1 = 10, carry the 1 from the 10 to the hundreds place.

   4,895 rounded to the nearest ten is 4,900.

Practice Exercises 1.1C:

1. Round 38,762 to the nearest hundred.

   38,800

2. Round 84,952 to the nearest thousand.

   85,000

3. Round 84,952 to the nearest hundred.

   85,000

4. Round 8,903,991 to the nearest ten.

   8,903,990

5. Round 8,903,991 to the nearest ten-thousand.

   8,900,000

Answers

Practice Exercises 1.1A:

1.

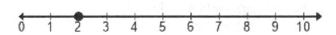

2.

8 is 6 units to the right of 2.

3. 1,049 > 109
4. 4,610 > 4,061
5. 918 < 8,978
6. 313, 1,111, 1,331, 3,113, 3,311
7. 13, 48, 72, 84, 93

Practice Exercises 1.1B:

1. The place value of the digit 0 is ten-thousands.
2. three billion, forty-nine million, two hundred forty-six
3. twelve billion, three hundred forty-five million, six hundred seventy-eight thousand, nine hundred
4. 7,000,008
5. 82,480,315,072
6. $50,000,000 + 8,000,000 + 30,000 + 200 + 40 + 1$
7. $7,000 + 200 + 40 + 5$

Practice Exercises 1.1C:

1. 38,800
2. 85,000
3. 85,000
4. 8,903,990
5. 8,900,000

Module 1: Whole Numbers
Objective 1.2A: Add Whole Numbers

Addition is the process of finding the total of two or more numbers.

To add large numbers, begin by arranging the numbers vertically, keeping digits of the same place value in the same column. When the sum of the numbers in a column exceeds 9, addition involves "carrying."

Example:

What is the sum of 487 and 369?

$$
\begin{array}{r}
{\scriptstyle 1} \\
4\ 8\ 7 \\
+\ 3\ 6\ 9 \\
\hline
6
\end{array}
$$

Add the ones column. $7 + 9 = 16$. Write the 6 in the ones column and carry the 1 ten to the tens column.

$$
\begin{array}{r}
{\scriptstyle 1}\ {\scriptstyle 1} \\
4\ 8\ 7 \\
+\ 3\ 6\ 9 \\
\hline
5\ 6
\end{array}
$$

Add the tens column. $1 + 8 + 6 = 15$. Write the 5 in the tens column and carry the 1 hundred to the hundreds column.

$$
\begin{array}{r}
{\scriptstyle 1}\ {\scriptstyle 1} \\
4\ 8\ 7 \\
+\ 3\ 6\ 9 \\
\hline
8\ 5\ 6
\end{array}
$$

Add the hundreds column. $1 + 4 + 3 = 8$. Write the 8 in the hundreds column.

The sum is 856.

A letter of the alphabet can be used in mathematics to stand for a number. Such a letter is called a **variable**. A mathematical expression that contains one or more variables is a **variable expression**. Replacing the variables in a variable expression with numbers and then simplifying the numerical expression is called **evaluating the variable expression**.

Example:

Evaluate $x + y + z$ for $x = 1,435$, $y = 2,982$, and $z = 8,648$.

Solution:

$$x + y + z$$
$$1,435 + 2,982 + 8,648$$

```
  2   1   1
      1   4   3   5
      2   9   8   2
+     8   6   4   8
─────────────────────
  1   3   0   6   5
```

Practice Exercises 1.2A:

1. Add:

```
    1   1   1
    6   7,  9   7   5
+   3   8,  8   9   2
─────────────────────
  1 0 6 8 6 7
```

2. Add: $54,458 + 8,192 + 903$

```
  1   1   1
5 4, 4 5 8
    8 1 9 2
        9 0 3
─────────────
  5 9 5 5 3
```

3. Add: $5,278 + 116 + 1,097$

```
    1   1
5 2 7 8
1 0 9 7
    1 1 6
─────────
  6 4 8 2
```

4. What is 784 more than 9,763?

$$\begin{array}{r} 9763 \\ 784 \\ \hline 10547 \end{array}$$

5. What is 56,801 increased by 3,299?

$$\begin{array}{r} 56801 \\ 3299 \\ \hline 60100 \end{array}$$

6. Find the total of 3,128 and 156.

$$\begin{array}{r} 3128 \\ 156 \\ \hline 3284 \end{array}$$

7. Evaluate $a+b$ for $a=18,196$ and $b=83,841$.

$$\begin{array}{r} 83841 \\ 18196 \\ \hline 102037 \end{array}$$

8. Evaluate $m+n+p$ for $m=7,777$, $n=888$, and $p=99$.

$$\begin{array}{r} 7777 \\ 888 \\ 99 \\ \hline 8764 \end{array}$$

Objective 1.2B: Subtract Whole Numbers

Subtraction is the process of finding the difference between two numbers. To subtract large numbers, begin by arranging the numbers vertically, keeping digits of the same place value in the same column. Then subtract the numbers in each column. When the lower digit is larger than the upper digit, subtraction involves "borrowing."

Examples:

1. Subtract: $843 - 217$
    Solution:

$\begin{array}{ccc} & {}^{3+1} & \\ 8 & \not{4} & 3 \\ - \ 2 & 1 & 7 \\ \hline \end{array}$	$\begin{array}{ccc} & {}^{3} & {}^{10} \\ 8 & \not{4} & 3 \\ - \ 2 & 1 & 7 \\ \hline \end{array}$	$\begin{array}{ccc} & {}^{3} & {}^{13} \\ 8 & \not{4} & \not{3} \\ - \ 2 & 1 & 7 \\ \hline \end{array}$	$\begin{array}{ccc} & {}^{3} & {}^{13} \\ 8 & \not{4} & \not{3} \\ - \ 2 & 1 & 7 \\ \hline 6 & 2 & 6 \end{array}$
$7 > 3$. Borrowing is necessary. 4 tens = 3 tens + 1 ten	Borrow 1 ten from the tens column and write 10 in the ones column.	Add the borrowed 10 to 3.	Subtract the numbers in each column.

Check: $626 + 217 = 843$

2. Subtract: $30,129 - 14,678$
    Solution:

$\begin{array}{cccccc} & {}^{2} & {}^{10} & & & \\ & \not{3} & \not{0}, & 1 & 2 & 9 \\ - & 1 & 4, & 6 & 7 & 8 \\ \hline \end{array}$	Borrow 1 ten-thousand from the ten-thousands column and write 10 in the thousands column.
$\begin{array}{cccccc} & {}^{2} & {}^{9}_{\not{10}} & {}^{11} & & \\ & \not{3} & \not{0}, & \not{1} & 2 & 9 \\ - & 1 & 4, & 6 & 7 & 8 \\ \hline \end{array}$	Borrow 1 thousand from the thousands column and write 11 in the hundreds column.
$\begin{array}{cccccc} & {}^{2} & {}^{9}_{\not{10}} & {}^{10}_{\not{11}} & {}^{12} & \\ & \not{3} & \not{0}, & \not{1} & \not{2} & 9 \\ - & 1 & 4, & 6 & 7 & 8 \\ \hline & 1 & 5, & 4 & 5 & 1 \end{array}$	Borrow 1 from the hundreds column and write 12 in the tens column.

Check: $15,451 + 14,678 = 30,129$

3. Evaluate $x - 4521$ for $x = 7,018$.

Solution:
$$x - 4521$$
$$7018 - 4521$$

$$
\begin{array}{r}
{}^6\!\!\not{7}\ {}^9_{\not{10}}\,0\ {}^{11}\!\!\not{1}\ 8 \\
-\ 4\ 5\ 2\ 1 \\
\hline
2\ 4\ 9\ 7
\end{array}
$$

Practice Exercises 1.2B:

1. Subtract: $58,426 - 37,592$

$$
\begin{array}{r}
5\,8\ {}^{1}\!4\,{}^{7}\!2\,{}^{13}\!6 \\
-\ 3\,7\ 5\,9\,2 \\
\hline
2\,0\ 8\,3\,4
\end{array}
$$

2. Subtract: $100,201 - 72,988$

$$
\begin{array}{r}
{}^{0}\!1\,{}^{9}\!0\,{}^{9}\!0\ {}^{11}\!2\,0\,{}^{9}\!1 \\
-\ \ 7\,2\,9\,8\,8 \\
\hline
2\,7\,2\,1\,3
\end{array}
$$

3. Subtract: $87,432 - 19,856$

$$
\begin{array}{r}
{}^{9}\!8\,{}^{16}\!7,\,{}^{13}\!4\,{}^{12}\!3\,{}^{12}\!2 \\
-\ 1\,9\ 8\,5\,6 \\
\hline
8\ 7\,5\,7\,6
\end{array}
$$

4. Subtract 381 from 1,024.

$$
\begin{array}{r}
{}^{0}\!1\,{}^{9}\!0\,{}^{12}\!2\,4 \\
-\ \ 3\,8\,1 \\
\hline
6\,4\,3
\end{array}
$$

5. What is 1,982 decreased by 898?

$$
\begin{array}{r}
1982 \\
-\ 898 \\
\hline
1084
\end{array}
$$

6. Evaluate $m - n$ when $m = 13,023$ and $n = 5,142$.

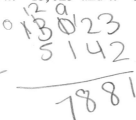

$$
\begin{array}{r}
13023 \\
-\ 5142 \\
\hline
7881
\end{array}
$$

7. Evaluate $x - y$ for $x = 923,456$ and $y = 765,567$

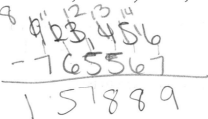

$$
\begin{array}{r}
923,456 \\
-\ 765567 \\
\hline
157889
\end{array}
$$

Objective 1.2C: Solve application problems

Some application problems can be solved with addition or subtraction.

Examples:

The diagram shows the projected numbers of electric cars sold in the United States for six years.

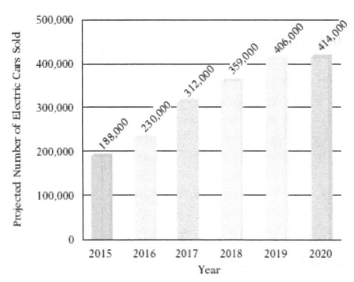

1. What is the total projected number of electric cars sold in 2015 and 2016?

2. How many more electric cars are projected to be sold in 2019 than in 2018?

Solutions:

1. Finding a total requires addition:

    ```
      1 8 8, 0 0 0
    + 2 3 0, 0 0 0
    ---------------
      4 1 8, 0 0 0
    ```

    It is projected that 418,000 will be sold in 2015 and 2016.

2. Finding a difference requires subtraction:

$$
\begin{array}{r}
\overset{3}{\cancel{4}}\ \overset{\overset{9}{\cancel{10}}}{\cancel{0}}\ \overset{16}{\cancel{6}}, 0\ 0\ 0 \\
-\ 3\ 5\ 9, 0\ 0\ 0 \\
\hline
4\ 7, 0\ 0\ 0
\end{array}
$$

It is projected that 47,000 more electric cars will be sold in 2019 than 2018.

Practice Exercises 1.2C:

The U.S. Census Bureau divides the United States into four regions, and it provides the following populations of those regions in 2013:

Region	Population
Northeast	55,943,073
Midwest	67,547,890
South	118,383,453
West	74,254,423

1. Estimate the combined population of the Northeast and Midwest regions by first rounding each region's population to the nearest million.

2. Find the total population of the Midwest, South, and West regions.

3. By how much does the population of the South exceed the population of the Midwest?

4. A rectangle has a length of 24 feet and a width of 5 feet. Find the perimeter of the rectangle.

5. The computer system you would like to purchase includes an operating system at $830, a monitor that costs $245, a keyboard priced at $175 and a printer that costs $395. What is the price of the computer system?

Answers

Practice Exercises 1.2A:

1. 106,867
2. 63,553
3. 6,491
4. 10,547
5. 60,100
6. 3,284
7. 102,037
8. 8,764

Practice Exercises 1.2B:

1. 20,834
2. 27,213
3. 67,576
4. 643
5. 1,084
6. 7,881
7. 157,889

Practice Exercises 1.2C:

1. 124,000,000
2. 260,185,766
3. 50,835,563
4. 58 feet
5. $1,645

Module 1: Whole Numbers
Objective 1.3A: Multiply whole numbers

Multiplication is used to find the total number of objects in several groups when each group contains the same number of objects. The **multiplicand** is the number of objects in each group (12 eggs in each carton); the **multiplier** is the number of groups (6 cartons); the **product** is the total number of objects (72 eggs). A **factor** is one of the numbers that are multiplied to obtain a product. 12 and 6 are factors of 72.

Examples:

1. Find the product of 946 and 7.

   Solution:

   $$
   \begin{array}{r}
   \overset{6}{}\overset{4}{} \\
   9\ 8\ 6 \\
   \times \qquad 7 \\
   \hline
   6\ 9\ 0\ 2
   \end{array}
   $$

   * $6 \times 7 = 42$
   * Write the 2 in the ones column. Carry the 4 to the tens column.
   * $7 \times 8 = 56, 56 + 4 = 60$
   * $7 \times 9 = 63, 63 + 6 = 69$

2. Find 413 multiplied by 794.

   Solution:

   $$
   \begin{array}{r}
   4\ 1\ 3 \\
   \times \qquad 7\ 9\ 4 \\
   \hline
   1\ 6\ 5\ 2 \\
   3\ 7\ 1\ 7 \\
   2\ 8\ 9\ 1 \\
   \hline
   3\ 2\ 7{,}9\ 2\ 2
   \end{array}
   $$

   * $4 \times 413 = 1{,}652$
   * $9 \times 413 = 3{,}717$
   * $7 \times 413 = 2{,}891$

3. Evaluate xyz for $x = 74$, $y = 5$, and $z = 22$.

   Solution:

   xyz

   $74 \cdot 5 \cdot 22$

   $370 \cdot 22$

   $8{,}140$

Practice Exercises 1.3A:

1. Multiply: 789×96

2. Multiply: $479 \cdot 892$

3. What is twice the product of 89 and 57?

4. Multiply 604 by 378.

5. Multiply 8,900 by 27.

6. Evaluate abc when $a = 4$, $b = 85$, and $c = 80$.

Objective 1.3B: Divide whole numbers

Division is used to separate objects into equal groups. If 30 objects were split among 5 containers (5 groups), then each container (group) would have 6 objects.

$$\begin{array}{r} 6 \\ 5\overline{)30} \end{array}$$

30 is the **dividend**. 5 is the **divisor**. 6 is the **quotient**.

Quotient $\times$ divisor $=$ dividend: $6 \times 5 = 30$

Division by zero is not allowed.

Examples:

1. Divide: $1{,}734 \div 6$

Solution:

$$\begin{array}{r} 2 \\ 6\overline{)1\ 7\ 3\ 4} \\ -1\ 2 \\ \hline 5\ 3 \end{array}$$

* Think $6\overline{)17}$. Place 2 in the quotient.
* Multiply 2×6.
* Subtract $17 - 12 = 5$. Bring down the 3.

$$\begin{array}{r} 2\ 8 \\ 6\overline{)1\ 7\ 3\ 4} \\ -1\ 2 \\ \hline 5\ 3 \\ -4\ 8 \\ \hline 5\ 4 \end{array}$$

* Think $6\overline{)53}$. Place 8 in the quotient.
* Multiply 8×6.
* Subtract $53 - 48 = 5$. Bring down the 4.

$$\begin{array}{r} 2\ 8\ 9 \\ 6\overline{)1\ 7\ 3\ 4} \\ -1\ 2 \\ \hline 5\ 3 \\ -4\ 8 \\ \hline 5\ 4 \\ -5\ 4 \\ \hline 0 \end{array}$$

* Think $6\overline{)54}$. Place 9 in the quotient.
* Multiply 9×6.
* Subtract $54 - 54 = 0$.

2. Find 7,068 divided by 23.

    Solution:

$$
\begin{array}{r}
3\;\;0\;\;7\ \text{r}\,7 \\
23\overline{)\;7\;\;0\;\;6\;\;8} \\
-\ 6\;\;9 \\
\hline
1\;\;6 \\
-0 \\
\hline
1\;\;6\;\;8 \\
-\ 1\;\;6\;\;1 \\
\hline
7
\end{array}
$$

* Think $23\overline{)16}^{\,0}$. Place 0 in the quotient.
* Multiply 0×23.
* Subtract $16 - 0$. Bring down the 8.

3. Evaluate $\dfrac{x}{42}$ for $x = 2{,}478$.

    Solution:

$$
\begin{array}{r}
5\;\;9 \\
42\overline{)\;2\;\;4\;\;7\;\;8} \\
-\ 2\;\;1\;\;0 \\
\hline
3\;\;7\;\;8 \\
-\ 3\;\;7\;\;8 \\
\hline
0
\end{array}
$$

Practice Exercises 1.3B:

1. Divide: $7{,}227 \div 9$

2. Divide: $26{,}031 \div 54$

3. What is the quotient of 6,381 and 9?

4. Evaluate $x \div y$ for $x = 59{,}200$ and $y = 8$.

5. Evaluate $\dfrac{x}{y}$ when $x = 0$ and $y = 19$.

Objective 1.3C: Solve application problems

Some application problems can be solved with multiplication or division.

Examples:

1. A moving company charges a $120 trip fee plus $125 per hour. What is the cost of a move that takes 3 hours?

    Solution:

    Since the move takes 3 hours at a rate of $125 per hour, multiply 125×3.

    ```
      1   2   5
    ×         3
    ───────────
      3   7   5
    ```

    The trip fee is $120. Add this to the $375.

    $375 + 120 = 495$.

    The move costs $495.

2. A used car costs $6,328. If you put $1,000 down and will pay the remaining balance over 3 years (36 months), what is your monthly payment?

    Solution:

    Subtract the $1,000 down payment from $6,328:

    ```
        6   3   2   8
    −   1   0   0   0
    ─────────────────
        5   3   2   8
    ```

    Divide $5,328 by 36:

    ```
                1   4   8
        ┌──────────────────
    36) │   5   3   2   8
        │ − 3   6
        │ ─────────
        │   1   7   2
        │ − 1   4   4
        │ ─────────
        │       2   8   8
        │     − 2   8   8
        │     ─────────
        │             0
    ```

    The monthly payment is $148.

Practice Exercises 1.3C:

1. Suppose that you will pay a mortgage of $1,324 per month for 30 years. What is the total amount you will pay?

2. An office area of 4,144 square feet is to have a 400-square-foot break room. The remaining area is reserved for 78 cubicles. What will the area of each cubicle be?

3. College of the Learned charges a $300 registration fee and $200 per credit hour. College of the Scholars charges a $150 registration fee and $220 per credit hour. Which college is more expensive for a 12-credit-hour load? By how much?

4. Find a.) the perimeter, and b.) the area of a rectangle with a length of 24 meters and a width of 15 meters.

Answers:

Practice Exercises 1.3A:

1. 75,744
2. 427,268
3. 10,146
4. 228,312
5. 240,300
6. 27,200

Practice Exercises 1.3B:

1. 803
2. 482.$\overline{3}$
3. 709
4. 7,400
5. 0

Practice Exercises 1.3C:

1. You will pay $476,640.
2. The area of each cubicle will be 48 square feet.
3. College of the Scholars is more expensive by $90 ($2,790 vs. $2,700).
4. a.) 78 meters b.) 360 square meters

Objective 1.4A: Simplify expressions that contain exponents

Repeated multiplication of the same factor can be written in two ways:

$$3 \cdot 3 \cdot 3 \cdot 3 \cdot 3 \cdot 3 \text{ or } 3^6$$

The expression 3^6 is in **exponential form**. The **exponent**, 6, indicates how many times the **base**, 3, occurs as a factor in the multiplication.

Examples:

1. Write $5 \cdot 5 \cdot 5 \cdot 5 \cdot 7 \cdot 7 \cdot 7$ in exponential form.

   Solution:

   $$5 \cdot 5 \cdot 5 \cdot 5 \cdot 7 \cdot 7 \cdot 7 = 5^4 \cdot 7^3$$

2. Evaluate 2^5.

   $$2^5 = 2 \cdot 2 \cdot 2 \cdot 2 \cdot 2 = 4 \cdot 2 \cdot 2 \cdot 2 = 8 \cdot 2 \cdot 2 = 16 \cdot 2 = 32$$

3. Evaluate 10^5.

   $$10^5 = 100,000$$

   (The exponent on 10 is 5. There are 5 zeros in 100,000.)

4. Evaluate ab^3 for $a = 3$ and $b = 5$.

   $$ab^3 = 3 \cdot 5^3 = 3 \cdot (5 \cdot 5 \cdot 5) = 3 \cdot 125 = 375$$

Practice Exercises 1.1A:

1. Write $3 \cdot 3 \cdot 5 \cdot 5 \cdot 5 \cdot 5 \cdot 5 \cdot 5 \cdot 7$ in exponential form.

2. Write $x \cdot x \cdot x \cdot x \cdot y \cdot y \cdot z \cdot z \cdot z$ in exponential form.

3. Evaluate 3^4.

4. Evaluate 4^3.

5. Evaluate $2^4 \cdot 5^2$.

6. Evaluate $7^2 \cdot 3^3$.

7. Evaluate $x^3 y^2$ when $x = 3$ and $y = 8$.

8. Evaluate $a^4 b^5$ when $a = 2$ and $b = 3$

Objective 1.4B: Use the Order of Operations Agreement to simplify expressions

To prevent there being more than one answer to the same problem, an Order of Operations Agreement is followed.

The Order of Operations Agreement

Step 1: Do all operations inside parentheses.

Step 2: Simplify any numerical expressions containing exponents.

Step 3: Do multiplication and division as they occur from left to right.

Step 4: Do addition and subtraction as they occur from left to right.

Examples:

1. Simplify: $42 - 3(4 + 2) \div 3^2$

    Solution:

$42 - 3(4 + 2) \div 3^2$	$= 42 - 3(6) \div 3^2$	Perform operations inside parentheses.
	$= 42 - 3(6) \div 9$	Simplify expresions with exponents.
	$= 42 - 18 \div 9$	Multiply or divide from left to right.
	$= 42 - 2$	Perform operations inside parentheses.
	$= 40$	Subtract.

2. Evaluate $x \div y \cdot z^2$ for $x = 12$, $y = 3$, and $z = 2$.

    Solution:

$x \div y \cdot z^2$	$= 12 \div 3 \cdot 2^2$	Simplify expressions with exponents.
	$= 12 \div 3 \cdot 4$	Multiply or divide from left to right.
	$= 4 \cdot 4$	
	$= 16$	

Practice Exercises 1.4B:

1. Simplify: $6 \cdot 4 + 5$

2. Simplify: $4^2 - 3$

3. Simplify: $25 + 2(14 - 6) \div 4^2$

4. Simplify: $17 - 2^3 \div 4 + 4$

5. Evaluate $x \div 2(y-z)^2$ for $x = 36$, $y = 5$, and $z = 2$.

6. Evaluate $a \div b^2 - 3c$ for $a = 128$, $b = 8$, and $c = 0$.

Answers:

Practice Exercises 1.4A:

1. $3^2 \cdot 5^6 \cdot 7$
2. $x^4 y^2 z^3$
3. 81
4. 64
5. 400
6. 1,323
7. 1,728
8. 3,888

Practice Exercises 1.4B:

1. 29
2. 13
3. 26
4. 19
5. 162
6. 2

Module 2: Integers
Objective 2.1A – Use inequality symbols with integers

Mathematicians place objects with similar properties in groups called sets. A **set** is a collection of objects. The objects in a set are called the **elements of the set**.

The **roster method** of writing a set encloses a list of the elements in braces.

The set of **natural numbers** is the set $\{1, 2, 3, 4, ...\}$.

The set of **integers** is the set $\{..., -3, -2, -1, 0, 1, 2, 3, ...\}$.

Each integer can be shown on a number line. The integers to the left of zero on the number line are called **negative integers**. The integers to the right of zero are called **positive integers**, or natural numbers. Zero is neither a positive nor a negative integer.

The **graph** of an integer is shown by placing a heavy dot on the number line directly above the number.

The graphs of –4 and 3 are shown on the number line below.

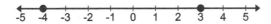

In mathematics, a letter of the alphabet can be used to stand for a number. Such a letter is called a **variable**.

If a and b are two numbers and a is to the left of b on the number line, then a **is less than** b. This is written $a < b$.

Example:

> On the number line above, –4 is to the left of 3. So –4 is less than 3. This is written $-4 < 3$.

If a and b are two numbers and a is to the right of b on the number line, then a **is greater than** b. This is written $a > b$.

Example:

> On the number line above, 3 is to the right of –4. So 3 is greater than –4. This is written $3 > -4$.

There are also inequality symbols for **is less than or equal to (≤)** and **is greater than or equal to (≥).**

Examples:

1. Use the roster method to write the set of negative integers greater than −5.

   Solution: $A = \{-4, -3, -2, -1\}$

2. Use the roster method to write the set of positive integers less than or equal to 4.

   Solution: $A = \{1, 2, 3, 4\}$ −1

3. Which is the lesser number,   or −3?

   Solution:

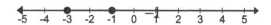

   Since −3 is to the left of   on the number line, −3 is the lesser number. $-3 < -1$.

4. Given $B = \{-5, -3, -1, 3, 5\}$, which elements of set B are less than or equal to −1?

   Solution: The elements −5 and −3 are less than or equal to −1.

Practice Exercises 2.1A

1. Place the correct inequality symbol, < or >, between the numbers below.

$$-13 \quad -15$$

2. Place the correct inequality symbol, < or >, between the numbers below.

$$-13 \quad 12$$

3. Place the correct inequality symbol, $<$ or $>$, between the numbers below.

$$0 \quad -1$$

4. Use the roster method to write the set of positive integers less than 6.

5. Use the roster method to write the set of negative integers greater than or equal to -7.

6. Which is the lesser temperature, $-8°F$ or $-5°F$?

7. Given $C = \{-8, -4, 0, 4, 8\}$, which elements of set C are greater than -4?

8. Given $C = \{-8, -4, 0, 4, 8\}$, which elements of set C are less than or equal to 0?

9. Graph the numbers -4 and 4 on the number line.

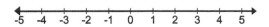

10. On the number line, which number is 7 units to the right of -5?

11. On the number line, which number is 3 units to the left of 2?

Objective 2.1B – Simplify expressions with absolute value

Two numbers that are the same distance from zero on the number line but are on opposite sides of zero are **opposite numbers**, or **opposites**. The opposite of a number is also called its **additive inverse**.

The negative sign can be read "the opposite of."

Examples:

1. $-(3) = -3$ The opposite of 3 is -3.
2. $-(-4) = 4$ The opposite of   is 4.
3. $-(8) = -8$ The additive inverse of 8 is

The **absolute value of a number** is its distance from zero on the number line. Therefore, the absolute value of a number is a positive number or zero. The symbol for absolute value is two vertical bars, $|\ |$.

The absolute value of a positive number is the number itself.

The absolute value of a negative number is the opposite of the number.

The absolute value of zero is zero.

Examples:

1. Evaluate $|-4|$.

   $|-4| = 4$

2. Evaluate $-|-5|$.

   $-|-5| = -5$

3. Evaluate $|21|$.

   $|21| = 21$

Practice Exercises 2.1B:

1. Evaluate $-(-2)$.

2. Find the additive inverse of -9.

3. Find the additive inverse of 8.

4. Find the opposite of 6.

5. Find the opposite of -17.

6. Evaluate $-(-23)$.

7. Evaluate $-|-23|$.

8. Evaluate $|-65|$.

9. Evaluate $-|3|$.

Place the correct symbol, <, =, or > between the two numbers.

10. $|-14|$ $\qquad$ $|17|$

11. $|-6|$ $\qquad$ $|6|$

12. $-|-25|$ $\qquad$ $-|-26|$

13. Write in order from smallest to largest:
$|2|, |-5|, -6, -|-20|$

Answers

Practice Exercises 2.1A:

1. $-13 > -15$
2. $-13 < 12$
3. $0 > -1$
4. $A = \{1, 2, 3, 4, 5\}$
5. $A = \{-7, -6, -5, -4, -3, -2, -1\}$
6. $-8°F$ is a lesser temperature than $-5°F$.
7. The elements of C that are greater than -4 are 0, 4, and 8.
8. The elements of C that are less than or equal to 0 are -8, -4, and 0.
9.
10. 2
11. -1

Practice Exercises 2.1B:

1. $-(-2) = 2$
2. The additive inverse of -9 is 9.
3. The additive inverse of 8 is -8.
4. The opposite of 6 is -6.
5. The opposite of -17 is 17.
6. $-(-23) = 23$
7. $-|-23| = -23$
8. $|-65| = 65$
9. $-|3| = -3$
10. $|-14| < |17|$
11. $|-6| = |6|$
12. $-|-25| > -|-26|$
13. $-|-20|, \ -6, \ |2|, \ |-5|$

Module 2: Integers
Objective 2.2A – Add integers

Addition is the process of finding the total of two numbers. The numbers being added are called **addends**. The total is called the **sum**.

To add two numbers with the same sign, add the absolute values of the numbers. Then attach the sign of the addends.

Example:

Add: $-34 + (-56)$

Solution: $|-34| = 34; \ |-56| = 56$

$34 + 56 = 90,$ and a negative sign must be attached.

$-34 + (-56) = -90$

To add two numbers with different signs, find the absolute value of each number. Subtract the smaller of the absolute values from the larger. Then attach the sign of the number with the larger absolute value.

Examples:

1. Add: $-34 + 56$

   Solution: $|-34| = 34; \ |56| = 56$

   $56 - 34 = 22,$ and the sign should be positive since (positive) 56 has the larger absolute value.

   $-34 + 56 = 22$

2. Add: $-54 + 19 + (-12)$

   Solution: $-54 + 19 + (-12) = -35 + (-12)$
   $$= -47$$

3. Find -87 increased by 21.

   Solution: $-87 + 21 = -66$

Practice Exercises 2.2A:

1. Add: $-7+(-4)$

2. Add: $-4+7$

3. Add: $45+(-45)$

4. Find the sum of $-13, 15,$ and 17.

5. What is 8 more than -5?

6. Add: $-12+15+(-18)$

7. Find -42 increased by 28.

8. Evaluate $x+y$, where $x=-30$ and $y=-22$.

9. Evaluate $-a+b$, where $a=-4$ and $b=-16$.

10. Evaluate $-x+y+z$, where $x=42$, $y=28$, and $z=-14$.

Objective 2.2B – Subtract integers

To **subtract** one number from another, add the opposite of the second number to the first number.

Examples:

1. Subtract: $-18-15$

    Solution: $-18-15 = -18+(-15) = -33$

2. Subtract: $-18-(-15)$

    Solution: $-18-(-15) = -18+15 = -3$

3. Subtract: $18-52$

    Solution: $18-52 = 18+(-52)$

    $\qquad\qquad |18| = 18;\ |-52| = 52$

    $\qquad\qquad 52-18 = 34,$ and a negative sign must be attached.

    $\qquad\qquad 18-52 = -34$

4. Find the difference between -5 and 6.

    Solution: The difference between -5 and 6 is $-5-6$.
    $\qquad\qquad -5-6 = -5+(-6) = -11$

5. Subtract -32 from -51.

    Solution: $-51-(-32) = -51+32 = -19$

Practice Exercises 2.2B:

1. Subtract: $-7-(-4)$

2. Subtract: $-4-7$

3. Subtract: $45-(-45)$

4. Find the difference of -13 and 17.

5. What is 8 decreased by -5?

6. Subtract: $-12-15-(-18)$

7. What is 28 less than -42?

8. Evaluate $x - y$, when $x = -8$ and $y = -42$.

9. Evaluate $-a - b + c$, when $a = 3$, $b = -4$, and $c = -9$.

10. Evaluate $-x + y - z$, where $x = -12$, $y = -7$, and $z = 3$.

Objective 2.2C – Solve application problems

Some application problems can be solved with addition or subtraction.

Examples:

Chemical Element	Boiling Point	Melting Point
Mercury	357	-39
Radon	-62	-71
Xenon	-108	-112

The table above shows the boiling point and the melting point, in degrees Celsius, of three chemical elements.

1. What is the difference between the boiling point of mercury and the melting point of mercury?

   Solution: Subtract the melting point of mercury from the boiling point.
   $$357 - (-39) = 357 + 39 = 396$$

2. What is the difference between the boiling point of xenon and the melting point of xenon?

   Solution: Subtract the melting point of xenon from the boiling point.
   $$-108 - (-112) = -108 + 112 = -4$$

3. Find the absolute value of the difference between the two lowest melting points in the table.

   Solution: The two lowest melting points are -112 and -71.

   $$\begin{aligned}|-112 - (-71)| &= |-112 + 71| \\ &= |42| \\ &= 42\end{aligned}$$

Practice Exercises 2.2C:

The table below shows the average temperatures at different cruising altitudes for airplanes.

Cruising Altitude	Average Temperature
12,000 ft	16°
20,000 ft	−12°
30,000 ft	−48°
40,000 ft	−70°
50,000 ft	−70°

1. What is the difference between the average temperatures at 12,000 ft and at 40,000 ft?

2. What is the difference between the average temperatures at 40,000 ft and 50,000 ft?

3. How much colder is the average temperature at 30,000 ft than at 12,000 ft?

4. Find the temperature after a rise of $9°C$ from $-9°C$.

5. Find the temperature after a decrease of $10°F$ from $3°F$.

6. Find the temperature after a decrease of $24°F$ from $52°F$.

Answers

Practice Exercises 2.2A:

1. −11
2. 3
3. 0
4. 19
5. 3
6. −15
7. −14
8. −52
9. −12
10. −28

Practice Exercises 2.2B:

1. −3
2. −11
3. 90
4. −30
5. 13
6. −9
7. −70
8. 34
9. −8
10. 2

Practice Exercises 2.2C:

1. 86°
2. 0°
3. 64°
4. 0°C
5. −7°F
6. 28°F

Module 2: Integers
Objective 2.3A – Multiply integers

Several different symbols are used to indicate multiplication. The numbers being multiplied are called **factors**. The result is called the **product**. Note that when parentheses are used and there is no operation symbol, the operation is multiplication.

To multiply two numbers with the same sign, multiply the absolute values of the numbers. The product is positive.

To multiply two numbers with different signs, multiply the absolute values of the numbers. The product is negative.

Examples:

1. Multiply: $-15(8)$

   Solution: $-15(8) = -120$

   (The signs are different, so the product is negative.)

2. Multiply: $(-15)(-8)$

   Solution: $(-15)(-8) = 120$

   (The signs are the same, so the product is positive.)

3. Find the product of $-3, -4,$ and $-5.$

   Solution: $(-3)(-4)(-5) = 12(-5)$
   $$= -60$$

4. Find the product of $-3, -4, -5,$ and $-6.$

   Solution: $(-3)(-4)(-5)(-6) = 12(-5)(-6)$
   $$= (-60)(-6)$$
   $$= 360$$

Note that when there is an even number of negative factors, the product is positive. When there is an odd number of negative factors, the product is negative.

Practice Exercises 2.3A:

1. Multiply: $(-14)(7)$

2. Multiply: $14(-7)$

3. Multiply: $8(-9)(-10)$

4. Multiply: $(15)3(-5)$

5. Find the product of $-2, -4, -6,$ and $-8.$

6. Evaluate $-xy$, where $x=-3$ and $y=7$.

7. Evaluate $-abc$, where $a=-9$, $b=-7$, and $c=8$.

8. Evaluate xyz, where $x=-10$, $y=4$, and $z=0$.

9. Evaluate $-xyz$, where $x=4$, $y=-5$, and $z=-100$.

Objective 2.3B – Divide integers

To divide two numbers with the same sign, divide the absolute values of the numbers. The quotient is positive.

To divide two numbers with different signs, divide the absolute values of the numbers. The quotient is negative.

Examples:

1. Divide: $(-21) \div 7$

    Solution: $(-21) \div 7 = -3$

    (The signs are different, so the quotient is negative.)

2. Divide: $(-21) \div (-7)$

    Solution: $(-21) \div (-7) = 3$

    (The signs are the same, so the quotient is positive.)

3. Divide: $\dfrac{64}{-4}$; $\dfrac{-64}{4}$; $-\dfrac{64}{4}$

    Solution: $\dfrac{64}{-4} = -16$; $\dfrac{-64}{4} = -16$; $-\dfrac{64}{4} = -16$

Example 3 above suggests the following rule:

If a and b are integers, and $b \neq 0$, then $\dfrac{-a}{b} = \dfrac{a}{-b} = -\dfrac{a}{b}$.

The symbol $\neq$ means "is not equal to."

Some properties of division:

1. If $a \neq 0$, $\dfrac{0}{a} = 0$.   Zero divided by any number other than zero is zero.

2. If $a \neq 0$, $\dfrac{a}{a} = 1$.   Any number other than zero divided by itself is 1.

3. $\dfrac{a}{1} = a$ A number divided by 1 is the number.

4. $\dfrac{a}{0}$ is undefined.   Division by 0 is not defined.

Practice Exercises 2.3B:

1. Divide: $(-54) \div 6$

2. Divide: $(-54) \div (-6)$

3. Divide: $\dfrac{72}{-3}$

4. Divide: $0 \div (-28)$

5. What is the quotient of -94 and 2?

6. What is the ratio of -222 and -111?

7. Divide: $32 \div 0$

8. What is the quotient of $-32{,}000$ and 100?

9. What is -390 divided by -3?

10. What is the quotient of 655 and -5?

Objective 2.3C – Solve application problems

Some application problems can be solved with multiplication or division.

Examples:

1. The daily low temperatures (in degrees Celsius) for five days in Windsor, Ontario, Canada were $-7°, -3°, -3°, 1°$, and $2°$. Find the average low temperature.

    Solution: To find the **average** of a set of numbers, add the numbers and divide by the number of numbers.

    $$\frac{-7+(-3)+(-3)+1+2}{5} = \frac{-10}{5} = -2$$

    The average low temperature was $-2°$C.

2. A credit to an account can be represented with a positive number, and a debit from an account can be represented with a negative number. If 5 credits of \$30 and 8 debits of \$40 are processed for and against an account with an initial balance of \$340, what is the ending balance in the account?

    Solution: The ending balance can be determined by computing

    $$340 + 5(30) + 8(-40) = 340 + 150 + (-320)$$
    $$= 490 + (-320)$$
    $$= 170$$

    The ending balance in the account is \$170.

Practice Exercises 2.3C:

1. The daily high temperatures, in degrees Celsius, for a city in northern Canada were $-10°, -12°, -15°$, and $-11°$. Find the average daily high temperature for this period.

2. The daily low temperatures, in degrees Fahrenheit, for a city in Alaska were $-8°, -3°, 4°, 2°, 0°,$ and $-1°$. Find the average daily low temperature for this period.

3. To discourage random guessing on a multiple-choice exam, an instructor assigns 5 points for a correct answer, -2 points for an incorrect answer, and -1 points for leaving a question blank. What is the score for a student who had 15 correct answers, 6 incorrect answers, and 4 answers left blank?

4. To discourage random guessing on a multiple-choice exam, an instructor assigns 6 points for a correct answer, -4 points for an incorrect answer, and -2 points for leaving a question blank. What is the score for a student who had 16 correct answers, 7 incorrect answers, and 2 answers left blank?

5. During the first week in May a stock fell 11 points. Then the stock rose 6 points during the next week. Find the net change in the value of the stock for the two week period.

Answers

Practice Exercises 2.3A:

1. −98
2. −98
3. 720
4. −225
5. 384
6. 21
7. −504
8. 0
9. −2,000

Practice Exercises 2.3B:

1. −9
2. 9
3. −24
4. 0
5. −47
6. 2
7. Undefined
8. −320
9. 130
10. −131

Practice Exercises 2.3C:

1. The average daily high temperature for this period is $-12°C$.
2. The average daily low temperature for this period is $-1°F$.
3. The student's score is 59 points.
4. The student's score is 64 points.
5. The net change in the stock is -5 points.

Module 2: Integers
Objective 2.4A – Simplify expressions containing exponents

Repeated multiplication of the same factor can be written using an exponent. The **exponent** indicates how many times the factor, which is called the **base**, occurs in the multiplication. In the expression 3^6, 6 is the exponent and 3 is the base. $3^6 = 3 \cdot 3 \cdot 3 \cdot 3 \cdot 3 \cdot 3$. The expression 3^6 is in **exponential form**. The expression $3 \cdot 3 \cdot 3 \cdot 3 \cdot 3 \cdot 3$ is in **factored form**.

3^1 is read "3 to the first power" or just "3." Usually the exponent 1 is not written.

3^2 is read "3 to the second power" or "3 squared."

3^3 is read "3 to the third power" or "3 cubed."

To evaluate an exponential expression, write each factor as many times as indicated by the exponent. Then multiply.

Examples:

1. Evaluate $(-3)^4$.

    Solution: This is the fourth power of -3.
    $$(-3)^4 = (-3)(-3)(-3)(-3) = 81$$

2. Evaluate -3^4.

    Solution: This is the opposite of the fourth power of 3.
    $$-3^4 = -3 \cdot 3 \cdot 3 \cdot 3 = -81$$

3. Evaluate $(-1)^3 (-2)^5 (-3)^2$.

    Solution: $(-1)^3 (-2)^5 (-3)^2 = -1 \cdot (-32) \cdot 9$
    $$= 32 \cdot 9$$
    $$= 288$$

Practice Exercises 2.4A:

1. Evaluate -4^2.

2. Evaluate $(-4)^2$.

3. Evaluate $(-3)^5$.

4. Evaluate $(-3)^6$.

5. Evaluate $(-1)^8(-4)^3(-2)^2$.

6. Evaluate $(-1)^{11} \cdot 3^3 \cdot (-2)^4$

7. Evaluate x^3y^4, where $x = -2$ and $y = 3$.

8. Evaluate a^2bc^3, where $a = -1$, $b = -4$, and $c = 2$.

9. Evaluate $-x^4y$, where $x = -2$ and $y = -10$.

10. Evaluate $-a^3b^2$, where $a = -10$ and $b = 6$.

Objective 2.4B – Use the Order of Operations Agreement to simplify expressions

To prevent there being more than one answer to the same problem, an Order of Operations Agreement is followed.

The Order of Operations Agreement

Step 1: Perform operations inside grouping symbols. Grouping symbols include parentheses ( ), brackets [ ], braces { }, the absolute value symbol | |, and the fraction bar.

Step 2: Simplify exponential expressions.

Step 3: Do multiplication and division as they occur from left to right.

Step 4: Do addition and subtraction as they occur from left to right.

One or more of the steps listed above may not be needed to evaluate an expression. In that case, proceed to the next step in the Order of Operations Agreement.

Examples:

1. Evaluate $24 \div \left[-10 - (1-3) \right] + 2^2$.

   Solution: $24 \div \left[-10 - (1-3) \right] + 2^2 = 24 \div \left[-10 - (-2) \right] + 2^2$
   $$= 24 \div (-8) + 2^2$$
   $$= 24 \div (-8) + 4$$
   $$= -3 + 4$$
   $$= 1$$

Evaluate $5 - \dfrac{4-20}{2-(-2)^2} - 3$.

Solution: $5 - \dfrac{4-20}{2-(-2)^2} - 3 = 5 - \dfrac{-16}{2-4} - 3$
$$= 5 - \frac{-16}{-2} - 3$$
$$= 5 - 8 - 3$$
$$= -3 - 3$$
$$= -6$$

Practice Exercises 2.4B:

1. Evaluate $-5^2 + 5 \cdot 2$.

2. Evaluate $12 \div (-6) \cdot 2 + 4$.

3. Evaluate $-28 \div 4 - 2^2 - (-3)^3$.

4. Evaluate $12 \div (3-5)^2 \cdot (-2)$.

5. Evaluate $9 + \dfrac{(-3)^2 + 1}{-1 - 2^2} \div (-2)$.

6. Evaluate $-8^2 + 3\left[4 - (1-3)^2\right]$.

7. Evaluate $15 + (-3)^2 \div 3 - 20$.

8. Evaluate $4a - 3c$, where $a = -3$ and $c = 0$.

9. Evaluate $\dfrac{d - b}{a}$ where $a = -3$, $b = -4$, and $d = 2$.

10. Evaluate $(d - a)^2 + (c - b)^2$, where $a = -5$, $b = 2$, $c = -1$, and $d = 4$.

Answers

Practice Exercises 2.4A:

1. −16
2. 16
3. −243
4. 729
5. −256
6. −432
7. −648
8. −32
9. 160
10. 36,000

Practice Exercises 2.4B:

1. −15
2. 0
3. 16
4. −6
5. 10
6. −64
7. −2
8. −12
9. −2
10. 90

Module 3: Fractions

Objective 3.1A – Factor numbers and use the prime factorization of numbers

Natural number **factors** of a number divide that number evenly (there is no remainder).

The following rules are helpful in finding the factors of a number.

1. 2 is a factor of a number if the digit in the ones place of the number is 0, 2, 4, 6, or 8.
2. 3 is a factor of a number if the sum of the digits of the number is divisible by 3.
3. 4 is a factor of a number if the last two digits of the number are divisible by 4.
4. 5 is a factor of a number if the ones digit of the number is 0 or 5.

Example:

    Find all the factors of 48.

    Solution:

       $48 \div 1 = 48$ 1 and 48 are factors of 48.

       $48 \div 2 = 24$ 2 and 24 are factors of 48.

       $48 \div 3 = 16$ 3 and 16 are factors of 48.

       $48 \div 4 = 12$ 4 and 12 are factors of 48.

       $48 \div 5$ 5 will not divide 48 evenly.

       $48 \div 6 = 8$ 6 and 8 are factors of 48.

       $48 \div 7$ 7 will not divide 48 evenly.

       $48 \div 8 = 6$ 8 and 6 are factors of 48.

    The factors are repeating. All the factors of 48 have been found.

    The factors of 42 are 1, 2, 3, 4, 6, 8, 12, 16, 24, and 48.

A **prime number** is a natural number greater than 1 that has exactly two natural number factors, 1 and the number itself. If a number is not prime, it is a **composite number**. The **prime factorization** of a number is the expression of the number as a product of its prime factors.

Example:

Find the prime factorization of 90.

Solution:

$$\begin{array}{r} 45 \\ 2\overline{)90} \end{array}$$

$$\begin{array}{r} 15 \\ 3\overline{)45} \\ 2\overline{)90} \end{array}$$

$$\begin{array}{r} 5 \\ 3\overline{)15} \\ 3\overline{)45} \\ 2\overline{)90} \end{array}$$

90 is divisible by 2. Divide 90 by 2.

45 is not divisible by 2 but is divisible by 3. Divide 45 by 3.

15 is divisible by 3. Divide 5 by 3. The quotient, 5, is prime.

The prime factorization of 90 is $2 \cdot 3^2 \cdot 5$.

Finding the prime factorization of larger numbers can be more difficult. Try each prime number as a trial divisor. Stop when the square of the trial divisor is greater than the number being factored.

Example:

Find the prime factorization of 355.

Solution: 355 is not divisible by 2 or by 3 but is divisible by 5. Divide 355 by 5:

$$\begin{array}{r} 71 \\ 5\overline{)355} \end{array}$$

71 cannot be divided evenly by 7 or 11. Prime numbers greater than 11 need not be tried because $11^2 = 121$ and $121 > 71$.

The prime factorization of 355 is $5 \cdot 71$.

Practice Exercises 3.1A:

1. Find all the factors of 18.

2. Find all the factors of 96.

3. Find all the factors of 97.

4. Find all the factors of 60.

5. Find all the factors of 72.

6. Find the prime factorization of 108.

7. Find the prime factorization of 109.

8. Find the prime factorization of 2,520.

9. Find the prime factorization of 72.

10. Find the prime factorization of 120.

Objective 3.1B – Find the least common multiple (LCM)

The **multiples of a number** are the products of that number and the numbers 1, 2, 3, 4, 5, …

A number that is a multiple of two or more numbers is a **common multiple** of those numbers. The **least common multiple (LCM)** is the smallest common multiple of two or more numbers.

Listing the multiples of each number is one way to find the LCM. Another way to find the LCM uses the prime factorization of each number. Find the prime factorization of each number and write the factorization of each number in a table. Mark the greatest product in each column. The LCM is the product of the marked numbers.

Examples:

1. Find the LCM of 126 and 220.

   Solution:

	2	**3**	**5**	**7**	**11**
126 =	2	3·3		7	
220 =	2·2		5		11

   The LCM = $2 \cdot 2 \cdot 3 \cdot 3 \cdot 5 \cdot 7 \cdot 11 = 13{,}860$.

2. Find the LCM of 9, 21, and 28.

   Solution:

	2	**3**	**7**
9 =		3·3	
21 =		3	7
28 =	2·2		7

   The LCM = $2 \cdot 2 \cdot 3 \cdot 3 \cdot 7 = 252$.

Practice Exercises 3.1B:

1. Find the LCM of 6 and 9.

2. Find the LCM of 12 and 15.

3. Find the LCM of 16 and 32.

4. Find the LCM of 106 and 108.

5. Find the LCM of 8, 10, and 12.

6. Find the LCM of 14, 16, and 52.

7. Find the LCM of 4 ,7, and 21.

8. Find the LCM of 2, 7, and 11.

9. Find the LCM of 2, 16, and 32.

10. Find the LCM of 8, 27, and 36.

Objective 3.1C – Find the greatest common factor (GCF)

A number that is a factor of two or more numbers is a **common factor** of those numbers. The **greatest common factor (GCF)** is the largest common factor of two or more numbers.

Listing the factors of each number is one way to find the GCF. Another way to find the GCF is to use the prime factorization of each number. Find the prime factorization of each number and write the factorization of each number in a table. Mark the least product in each column that does not have a blank. The GCF is the product of the marked numbers.

Examples:

1. Find the GCF of 126 and 220.

    Solution:

	2	3	5	7	11
126 =	2	3·3		7	
220 =	2·2		5		11

    The GCF = 2.

2. Find the GCF of 98, 1,008, and 1,176.

    Solution:

	2	3	7
98 =	2		7·7
1,008 =	2·2·2·2	3·3	7
1,176 =	2·2·2	3	7·7

    The GCF = $2 \cdot 7 = 14$.

Practice Exercises 3.1C:

1. Find the GCF of 12 and 20.

2. Find the GCF of 75 and 375.

3. Find the GCF of 144 and 360.

4. Find the GCF of 60, 72, and 96.

5. Find the GCF of 52, 78, and 156.

6. Find the GCF of 90, 91, and 93.

7. Find the GCF of 10, 25, and 50.

8. Find the GCF of 16, 20, and 32.

9. Find the GCF of 18, 27, and 81.

10. Find the GCF of 28, 44, and 56.

Answers

Practice Exercises 3.1A:

1. 1, 2, 3, 6, 9, 18
2. 1, 2, 3, 4, 6, 8, 12, 16, 24, 32, 48, 96
3. 1, 97
4. 1, 2, 3, 4, 5, 6, 10, 12, 15, 20, 30, 60
5. 1, 2, 3, 4, 6, 8, 9, 12, 18, 24, 36, 72
6. $2^2 \cdot 3^3$
7. Prime
8. $2^3 \cdot 3^2 \cdot 5 \cdot 7$
9. $2^3 \cdot 3^2$
10. $2^3 \cdot 3 \cdot 5$

Practice Exercises 3.1B:

1. 18
2. 60
3. 32
4. 5,724
5. 120
6. 1,456
7. 84
8. 154
9. 32
10. 216

Practice Exercises 3.1C:

1. 4
2. 75
3. 72
4. 12
5. 26
6. 1
7. 5
8. 4
9. 9
10. 4

Module 3: Fractions
Objective 3.2A – Write a fraction that represents part of a whole

A **fraction** can represent the number of equal parts of a whole.

Example:

The shaded portion of the circle is represented by the fraction $\frac{2}{3}$. Two of the three parts of the circle (that is, two-thirds of it) are shaded.

$$\text{Fraction bar} \rightarrow \frac{2 \leftarrow \text{Numerator}}{3 \leftarrow \text{Denominator}}$$

A **proper fraction** is a fraction less than 1. The numerator of a proper fraction is smaller than the denominator.

A **mixed number** is a number greater than 1 with a whole-number part and a fractional part.

Example:

The shaded portion of the circles can be represented by the mixed number $2\frac{2}{3}$.

An **improper fraction** is a fraction greater than or equal to 1. The numerator of an improper fraction is greater than or equal to the denominator.

Example:

The shaded portion of the circles can be represented by the improper fraction $\frac{8}{3}$.

Practice Exercises 3.2A:

1. Express the shaded portion of the circle as a fraction.

2. Express the shaded portion of the circle as a fraction.

3. Express the shaded portion of the circles as a mixed number.

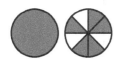

4. Express the shaded portion of the circles as a mixed number.

5. Express the shaded portion of the circles as an improper fraction.

Objective 3.2B – Write an improper fraction as a mixed number
or a whole number, and a mixed number as an improper
fraction

Note from the diagram that the mixed
number $1\frac{2}{5}$ and the improper fraction

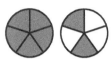

$\frac{7}{5}$ both represent the shaded portion of

the circles, so $1\frac{2}{5} = \frac{7}{5}$.

An improper fraction can be written as a mixed number or a whole
number.

Examples:

1. Write $\frac{7}{5}$ as a mixed number.

    Solution: Divide the numerator by the denominator.

    $$\begin{array}{r} 1 \\ 5{\overline{)7}} \\ \underline{-5} \\ 2 \end{array}$$

    To write the fractional part of the mixed number, write the
    remainder over the divisor.

    $$\frac{7}{5} = 1\frac{2}{5}$$

2. Write $\frac{27}{9}$ as a whole number.

    Solution: $\frac{27}{9} = 27 \div 9 = 3$

To write a mixed number as an improper fraction, multiply the
denominator of the fractional part by the whole-number part. The sum of
this product and the numerator of the fractional part is the numerator of
the improper fraction. The denominator remains the same.

Example:

Write $8\frac{3}{5}$ as an improper fraction.

Solution: $8\frac{3}{5} = \frac{(5 \times 8) + 3}{5} = \frac{40 + 3}{5} = \frac{43}{5}$

Practice Exercises 3.2B:

1. Write $\frac{19}{7}$ as a mixed or whole number.

2. Write $\frac{84}{6}$ as a mixed or whole number.

3. Write $\frac{21}{4}$ as an mixed or whole number.

4. Write $\frac{44}{3}$ as an mixed or whole number.

5. Write $5\dfrac{7}{9}$ as an improper fraction.

6. Write $12\dfrac{4}{13}$ as an improper fraction.

7. Write $6\dfrac{1}{7}$ as an improper fraction.

8. Write 12 as an improper fraction.

Answers

Practice Exercises 3.2A:

1. $\dfrac{5}{6}$

2. $\dfrac{3}{8}$

3. $1\dfrac{5}{8}$

4. $3\dfrac{3}{5}$

5. $\dfrac{18}{5}$

Practice Exercises 3.2B:

1. $2\dfrac{5}{7}$

2. 14

3. $5\dfrac{1}{4}$

4. $14\dfrac{2}{3}$

5. $\dfrac{52}{9}$

6. $\dfrac{160}{13}$

7. $\dfrac{43}{7}$

8. $\dfrac{12}{1}$

Module 3: Fractions
Objective 3.3A – Write a fraction in simplest form

Writing the **simplest form** of a fraction means writing it so that the numerator and denominator have no common factors other than 1.

The Multiplication Property of One can be used to write fractions in simplest form. Write the numerator and denominator of the given fraction as a product of factors. Write factors common to both the numerator and denominator as an improper fraction equivalent to 1.

$$\frac{9}{12} = \frac{3 \cdot 3}{2 \cdot 2 \cdot 3} = \frac{3}{2 \cdot 2} \cdot \frac{3}{3} = \frac{3}{2 \cdot 2} \cdot 1 = \frac{3}{4}$$

The process of eliminating common factors is displayed with slashes through the common factors as shown below:

$$\frac{9}{12} = \frac{3 \cdot \cancel{3}^{1}}{2 \cdot 2 \cdot \cancel{3}_{1}} = \frac{3}{2 \cdot 2} = \frac{3}{4}$$

To write a fraction in simplest form, eliminate the common factors. An improper fraction can be changed to a mixed number.

Examples:

1. Write $\dfrac{26}{39}$ in simplest form.

    Solution: $\dfrac{26}{39} = \dfrac{2 \cdot \cancel{13}^{1}}{3 \cdot \cancel{13}_{1}} = \dfrac{2}{3}$

2. Write $\dfrac{15}{45}$ in simplest form.

    Solution: $\dfrac{15}{45} = \dfrac{\cancel{3}^{1} \cdot \cancel{5}^{1}}{\cancel{3}_{1} \cdot 3 \cdot \cancel{5}_{1}} = \dfrac{1}{3}$

3. Write $\frac{10}{21}$ in simplest form.

Solution: $\frac{10}{21} = \frac{2 \cdot 5}{3 \cdot 7} = \frac{10}{21}$

$\frac{10}{21}$ is already in simplest form because there are no common factors in the numerator and denominator.

4. Write $\frac{26}{8}$ as a fraction in simplest form.

Solution: $\frac{26}{8} = \frac{\overset{1}{\cancel{2}} \cdot 13}{\underset{1}{\cancel{2}} \cdot 2 \cdot 2} = \frac{13}{4}$

Note that $\frac{13}{4}$ is a fraction in simplest form. This fraction is equivalent to the mixed number $3\frac{1}{4}$.

Practice Exercises 3.3A:

1. Write $\frac{32}{80}$ in simplest form.

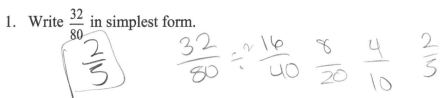

$\boxed{\dfrac{2}{5}}$ $\qquad \dfrac{32}{80} \div 2 \ \dfrac{16}{40} \quad \dfrac{8}{20} \quad \dfrac{4}{10} \quad \dfrac{2}{5}$

2. Write $\frac{9}{27}$ in simplest form.

$\dfrac{9}{27} \overset{\div 9}{} \qquad \boxed{\dfrac{1}{3}}$

3. Write $\frac{20}{27}$ in simplest form.

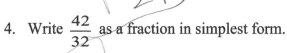

$\boxed{\dfrac{20}{27}}$

4. Write $\frac{42}{32}$ as a fraction in simplest form.

$\boxed{\dfrac{21}{16}} \qquad \dfrac{42}{32} \div 2 \quad \dfrac{21}{16}$

5. Write $\dfrac{45}{20}$ as a fraction in simplest form.

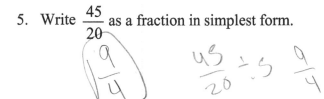

$\dfrac{9}{4}$

$\dfrac{45}{20} \div 5 \quad \dfrac{9}{4}$

6. Write $\dfrac{36}{8}$ as a mixed number in simplest form.

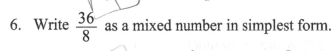

$\dfrac{9}{2}$

$\dfrac{36}{8} \div 4 \quad \dfrac{9}{2}$

7. Write $\dfrac{110}{6}$ as a mixed number in simplest form.

$\dfrac{55}{3}$

$\dfrac{110}{6} \div 2 \quad \dfrac{55}{3}$

Objective 3.3B – Find equivalent fractions by raising to higher terms

Equal fractions with different denominators are called **equivalent fractions**.

The Multiplication Property of One states that the product of a number and 1 is the number. This property can be used to write equivalent fractions.

Examples:

1. Write $\dfrac{2}{5}$ as an equivalent fraction with denominator 30.

    Solution:

    Divide the larger denominator by the smaller.

    $$30 \div 5 = 6$$

    Multiply the numerator and denominator of the given fraction by the quotient (6).

    $$\frac{2}{5} = \frac{2 \cdot 6}{5 \cdot 6} = \frac{12}{30}$$

    $\dfrac{2}{5}$ is equivalent to $\dfrac{12}{30}$.

2. Write $\dfrac{5}{8}$ as an equivalent fraction with denominator 96.

    Solution:

    $$96 \div 8 = 12$$

    $$\frac{5}{8} = \frac{5 \cdot 12}{8 \cdot 12} = \frac{60}{96}$$

    $\dfrac{5}{8}$ is equivalent to $\dfrac{60}{96}$.

3. Write 9 as a fraction with denominator 12.

   Solution:

   Write 9 as $\frac{9}{1}$.

   $12 \div 1 = 12$

   $\frac{9}{1} = \frac{9 \cdot 12}{1 \cdot 12} = \frac{108}{12}$

   9 is equivalent to $\frac{108}{12}$.

Practice Exercises 3.3B:

1. Write $\frac{5}{6}$ as an equivalent fraction with denominator 18.

   $\frac{5}{6}$ x3 / x3 $\boxed{\frac{15}{18}}$

2. Write $\frac{4}{7}$ as an equivalent fraction with denominator 28.

   $\frac{4}{7}$ x4 / x4 $\boxed{\frac{16}{28}}$

3. Write $\frac{11}{15}$ as an equivalent fraction with denominator 180.

   $\frac{11}{15}$ x12 / x12 $\boxed{\frac{132}{180}}$

4. Write 5 as a fraction with denominator 7.

   $\frac{5}{1}$ x7 / x7 $\boxed{\frac{35}{7}}$

5. Write 8 as an fraction with denominator 11.

$$\frac{8}{1} \begin{array}{c} \times 11 \\ \times 11 \end{array} \qquad \boxed{\frac{88}{11}}$$

6. Write $\frac{2}{3}$ as an equivalent fraction with numerator 16.

$$\frac{2}{3} \begin{array}{c} \times 8 \\ \times 8 \end{array} \qquad \boxed{\frac{16}{18}}$$

7. Write $\frac{1}{8}$ as an equivalent fraction with numerator 7.

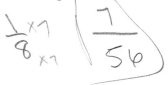

$$\frac{1}{8} \begin{array}{c} \times 7 \\ \times 7 \end{array} \qquad \boxed{\frac{7}{56}}$$

8. Write $\frac{5}{12}$ as an equivalent fraction with numerator 15.

$$\frac{5}{12} \begin{array}{c} \times 3 \\ \times 3 \end{array} \qquad \boxed{\frac{15}{36}}$$

Objective 3.3C – Identify the order relation between two fractions

Recall that whole numbers can be graphed as points on the number line. Fractions can also be graphed as points on the number line. The number line can be used to determine the order relation between two fractions. A fraction that appears to the left of a given fraction is less than the given fraction. A fraction that appears to the right of a given fraction is greater than the given fraction.

Example:

$$\frac{4}{7} < \frac{6}{7}$$

To find the order relation between two fractions with the same denominator, compare the numerators. The fraction that has the smaller numerator is the smaller fraction.

Example:

$$\frac{4}{7} < \frac{6}{7} \text{ since } 4 < 6.$$

When the denominators are different, begin by writing equivalent fractions with a common denominator; then compare the numerators.

Example:

Find the order relation between $\frac{5}{6}$ and $\frac{7}{9}$.

Solution: The LCD of the fractions is 18.

$$\frac{5}{6} = \frac{15}{18} \leftarrow \text{Larger numerator}$$

$$\frac{7}{9} = \frac{14}{18} \leftarrow \text{Smaller numerator}$$

$$\frac{5}{6} > \frac{7}{9} \text{ or } \frac{7}{9} < \frac{5}{6}$$

Practice Exercises 3.3C

1. Place the correct symbol, < or >, between the two numbers.

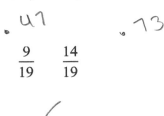

$$\frac{9}{19} \qquad \frac{14}{19}$$

$$<$$

2. Place the correct symbol, < or >, between the two numbers.

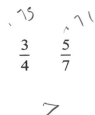

$$\frac{3}{4} \qquad \frac{5}{7}$$

$$>$$

3. Place the correct symbol, < or >, between the two numbers.

$$\frac{6}{11} \qquad \frac{8}{13}$$

$$<$$

4. Place the correct symbol, < or >, between the two numbers.

$$\frac{7}{15} \qquad \frac{11}{23}$$

$$<$$

5. Place the correct symbol, < or >, between the two numbers.

$$\frac{8}{5} \qquad \frac{10}{7}$$

>

6. Place the correct symbol, < or >, between the two numbers.

$$\frac{22}{4} \qquad \frac{35}{6}$$

<

Answers

Practice Exercises 3.3A:

1. $\dfrac{2}{5}$

2. $\dfrac{1}{3}$

3. $\dfrac{20}{27}$

4. $\dfrac{21}{16}$

5. $\dfrac{9}{4}$

6. $4\dfrac{1}{2}$

7. $18\dfrac{1}{3}$

Practice Exercises 3.3B:

1. $\dfrac{15}{18}$

2. $\dfrac{16}{28}$

3. $\dfrac{132}{180}$

4. $\dfrac{35}{7}$

5. $\dfrac{88}{11}$

6. $\dfrac{16}{24}$

7. $\dfrac{7}{56}$

8. $\dfrac{15}{36}$

Practice Exercises 3.3C:

1. $\dfrac{9}{19} < \dfrac{14}{19}$

2. $\dfrac{3}{4} > \dfrac{5}{7}$

3. $\dfrac{6}{11} < \dfrac{8}{13}$

4. $\dfrac{7}{15} < \dfrac{11}{23}$

5. $\dfrac{8}{5} > \dfrac{10}{7}$

6. $\dfrac{22}{4} < \dfrac{35}{6}$

Module 3: Fractions
Objective 3.4A – Multiply Fractions

To multiply two fractions, multiply the numerators and multiply the denominators.

Examples:

1. Multiply: $\dfrac{2}{3} \cdot \dfrac{6}{7}$

    Solution:

    $$\dfrac{2}{3} \cdot \dfrac{6}{7} = \dfrac{2 \cdot 6}{3 \cdot 7} \quad \text{(Multiply the numerators.)}$$
    $$\text{(Multiply the denominators.)}$$
    $$= \dfrac{2 \cdot 2 \cdot 3}{3 \cdot 7}$$
    $$= \dfrac{4}{7} \quad \text{(Write the product in simplest form.)}$$

2. Multiply: $\dfrac{2}{5}\left(\dfrac{7}{10}\right)\left(\dfrac{25}{7}\right)$

    Solution:

    $$\dfrac{2}{5}\left(\dfrac{7}{10}\right)\left(\dfrac{25}{7}\right) = \dfrac{2 \cdot 7 \cdot 25}{5 \cdot 10 \cdot 7}$$
    $$= \dfrac{2 \cdot 7 \cdot 5 \cdot 5}{5 \cdot 2 \cdot 5 \cdot 7}$$
    $$= \dfrac{1}{1} = 1$$

3. Find the product of 6 and $\dfrac{3}{4}$.

    Solution:

$$6 \cdot \frac{3}{4} = \frac{6}{1} \cdot \frac{3}{4}$$

$$= \frac{2 \cdot 3 \cdot 3}{2 \cdot 2}$$

$$= \frac{9}{2} = 4\frac{1}{2} \quad \text{(The answer can be written as an improper fraction or a mixed number.)}$$

4. Find the product of $2\frac{2}{3}$ and $5\frac{1}{4}$.

Solution:

$$2\frac{2}{3} \cdot 5\frac{1}{4} = \frac{8}{3} \cdot \frac{21}{4}$$

$$= \frac{2 \cdot 2 \cdot 2 \cdot 3 \cdot 7}{3 \cdot 2 \cdot 2}$$

$$= \frac{14}{1} = 14$$

5. Evaluate the variable expression xy for $x = 3\frac{5}{8}$ and $y = \frac{4}{5}$.

Solution:

$$xy$$

$$3\frac{5}{8} \cdot \frac{4}{5} = \frac{29}{8} \cdot \frac{4}{5}$$

$$= \frac{29 \cdot 2 \cdot 2}{2 \cdot 2 \cdot 2 \cdot 5}$$

$$= \frac{29}{10} = 2\frac{9}{10}$$

Practice Exercises 3.4A

1. Find the product of $\frac{5}{6}$ and $\frac{3}{10}$.

2. Multiply: $\dfrac{4}{7}\left(\dfrac{10}{11}\right)\left(\dfrac{14}{15}\right)$ = $\dfrac{40}{77} \cdot \dfrac{14}{15}$ = $\dfrac{560}{1155}$

$\dfrac{16}{33}$

3. What is the product of 8 and $\dfrac{9}{10}$?

$\dfrac{8}{1} \cdot \dfrac{9}{10} = \dfrac{72}{10}$ = $\dfrac{36}{5}$

4. What is the product of $5\dfrac{3}{5}$ and $3\dfrac{4}{7}$?

$\dfrac{28}{5} \cdot \dfrac{25}{7}$

$\dfrac{700}{35} = 20$

5. Evaluate ab for $a = 8\dfrac{1}{3}$ and $b = \dfrac{9}{10}$.

$\dfrac{25}{3} \cdot \dfrac{9}{10} = \dfrac{225}{30} = \dfrac{15}{2}$

6. Evaluate xyz for $x = 1\dfrac{2}{3}$, $y = \dfrac{4}{5}$, and $z = 6$.

$\dfrac{5}{3} \cdot \dfrac{4}{5} \cdot \dfrac{6}{1}$

$\dfrac{20}{13} \cdot \dfrac{6}{1} = \dfrac{120}{13}$

7. Find twice $3\frac{4}{5}$.

$$\frac{19}{5} \cdot \frac{19}{5} = \frac{361}{25}$$

8. Evaluate abc when $a = \frac{2}{9}$, $b = 3$, and $c = 4\frac{1}{10}$.

$$\frac{2}{9} \cdot \frac{3}{1} \cdot \frac{41}{10}$$

$$\frac{6}{9} \cdot \frac{41}{10} = \frac{246}{90} = \boxed{\frac{41}{15}}$$

Objective 3.4B – Divide fractions

The **reciprocal** of a fraction is that fraction with the numerator and denominator interchanged. The reciprocal of a number is also called the *multiplicative inverse* of the number. The product of a number and its reciprocal is 1.

Example: The reciprocal of $\frac{7}{9}$ is $\frac{9}{7}$, and $\frac{7}{9} \cdot \frac{9}{7} = 1$.

The process of interchanging the numerator and denominator of a fraction is called **inverting the fraction**. To find the reciprocal of a whole number, first rewrite the whole number as a fraction with a denominator of 1. Then invert the fraction.

Example: Find the reciprocal of 9.

Solution:

$9 = \frac{9}{1}$

Invert $\frac{9}{1}$ to get $\frac{1}{9}$.

The reciprocal of 9 is $\frac{1}{9}$.

Division is defined as multiplication by the reciprocal. Therefore, "divided by 9" is the same as "times $\frac{1}{9}$." Fractions are divided by making this substitution.

Examples:

1. Divide: $\frac{2}{3} \div \frac{6}{9}$

    Solution:

    $$\frac{2}{3} \div \frac{6}{9} = \frac{2}{3} \cdot \frac{9}{6}$$
    $$= \frac{2 \cdot 3 \cdot 3}{3 \cdot 2 \cdot 3}$$
    $$= 1$$

2. Find the quotient of $\dfrac{5}{6}$ and 3.

Solution:

$$\dfrac{5}{6} \div 3 = \dfrac{5}{6} \div \dfrac{3}{1}$$
$$= \dfrac{5}{6} \cdot \dfrac{1}{3}$$
$$= \dfrac{5}{2 \cdot 3 \cdot 3}$$
$$= \dfrac{5}{18}$$

3. Divide: $2\dfrac{3}{5} \div \dfrac{7}{10}$

Solution:

$$2\dfrac{3}{5} \div \dfrac{7}{10}$$
$$= \dfrac{13}{5} \div \dfrac{7}{10}$$
$$= \dfrac{13}{5} \cdot \dfrac{10}{7}$$
$$= \dfrac{13 \cdot 2 \cdot 5}{5 \cdot 7}$$
$$= \dfrac{26}{7} = 3\dfrac{5}{7}$$

4. Find the quotient of $3\dfrac{1}{2}$ and $1\dfrac{3}{4}$.

Solution:

$$3\frac{1}{2} \div 1\frac{3}{4} = \frac{7}{2} \div \frac{7}{4}$$

$$= \frac{7}{2} \cdot \frac{4}{7}$$

$$= \frac{7 \cdot 2 \cdot 2}{2 \cdot 7}$$

$$= 2$$

5. Evaluate $x \div y$ for $x = 5\frac{1}{3}$ and $y = \frac{2}{3}$.

Solution:

$$x \div y$$

$$5\frac{1}{3} \div \frac{2}{3} = \frac{16}{3} \div \frac{2}{3}$$

$$= \frac{16}{3} \cdot \frac{3}{2}$$

$$= \frac{2 \cdot 2 \cdot 2 \cdot 2 \cdot 3}{3 \cdot 2}$$

$$= 8$$

Practice Exercises 3.4B:

1. What is the quotient of $\frac{7}{8}$ and $\frac{3}{8}$?

$$\frac{7}{8} \cdot \frac{8}{3} = \boxed{\frac{7}{3}}$$

2. Find the quotient of $\frac{2}{5}$ and 4.

$$\frac{2}{5} \cdot \frac{1}{4} = \frac{2}{20} = \frac{1}{5}$$

3. Divide: $5\dfrac{1}{3} \div 6\dfrac{1}{3}$

$$\dfrac{16}{3} \cdot \dfrac{3}{19} \quad = \boxed{\dfrac{16}{19}}$$

4. Divide: $\dfrac{8}{17} \div \dfrac{3}{7}$

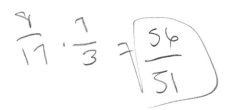

$$\dfrac{8}{17} \cdot \dfrac{7}{3} = \boxed{\dfrac{56}{51}}$$

5. Divide: $\dfrac{2}{3} \div 1\dfrac{7}{8}$

$$\dfrac{2}{3} \cdot \dfrac{8}{15} = \boxed{\dfrac{16}{45}}$$

6. Evaluate $a \div b$ for $a = 9$ and $b = 1\dfrac{5}{12}$.

$$\dfrac{9}{1} \div \dfrac{12}{17}$$

$$\dfrac{9}{1} \cdot \dfrac{12}{17} = \dfrac{108}{17} =$$

7. Evaluate $x \div y$ for $x = \dfrac{2}{5}$ and $y = \dfrac{8}{15}$.

$$\frac{2}{5} \cdot \frac{15}{8} = \frac{30}{40} = \frac{3}{4}$$

8. Evaluate $\dfrac{a}{b}$ for $a = 18$ and $b = 4\dfrac{1}{2}$.

$$\frac{18}{1} \div \frac{2}{9}$$

$$\frac{18}{1} \cdot \frac{2}{9} = \frac{36}{9} \quad \boxed{4}$$

Objective 3.4C – Solve application problems and use formulas

Some application problems can be solved with fraction multiplication and division.

Examples:

1. The formula for the area of a triangle is $A = \dfrac{1}{2}bh$, where A is the area of the triangle, b is the base, and h is the height. In $\triangle ABC$, if $\overline{AB}$ is the base, then the line segment from C that forms a right angle with the base is the height.

    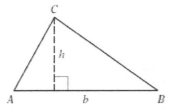

    Find the amount of material needed to make a decoration in the shape of a triangle with a base of 3 feet and a height of 2 feet.

    Solution: Use the formula for the area of a triangle, $A = \dfrac{1}{2}bh$, with $b = 3$ and $h = 2$.

    $$A = \frac{1}{2}bh$$
    $$A = \frac{1}{2}(3)(2)$$
    $$A = 3$$

    The area is 3 ft^2.

2. A 30-foot board is cut into pieces $2\dfrac{2}{3}$ ft long. What is the length of the remaining piece after as many $2\dfrac{2}{3}$-ft length as possible are cut?

Solution: The number of $2\frac{2}{3}$-ft lengths is found with division.

$$30 \div 2\frac{2}{3} = 30 \div \frac{8}{3} = \frac{30}{1} \cdot \frac{3}{8} = \frac{2 \cdot 3 \cdot 5 \cdot 3}{2 \cdot 2 \cdot 2} = \frac{45}{4} = 11\frac{1}{4}$$

11 full $2\frac{2}{3}$-ft lengths can be cut. The piece remaining is $\frac{1}{4}$ of a full $2\frac{2}{3}$-ft length. The length of the remaining piece is found with multiplication.

$$\frac{1}{4} \cdot 2\frac{2}{3} = \frac{1}{4} \cdot \frac{8}{3} = \frac{2 \cdot 2 \cdot 2}{2 \cdot 2 \cdot 3} = \frac{2}{3}$$

The remaining piece has length $\frac{2}{3}$ ft.

Practice Exercises 3.4C:

1. A triangular grazing area has length 90 meters and height 80 meters. What is its area?

2. A small-business website estimates that $\frac{4}{5}$ of a dollar is a reasonable estimate for the production cost per square foot of a billboard. If the area of a billboard is found by multiplying its length and its width, and if company wishes to advertise using a 14-foot by 50-foot billboard, then how much should it budget for billboard production?

3. An object can be assembled in $7\frac{1}{2}$ minutes. How many full objects can be assembled in 1 hour?

4. A project you are working on requires pieces of fabric $3\frac{1}{4}$-meters long. You have a piece of fabric that is 90 meters long. What is the length of the remaining piece?

5. Find the area of a triangle with $b = 3\frac{2}{5}$ feet and $h = 10$ feet.

6. Find the area of a rectangle with $L = 4\frac{1}{2}$ inches and $W = 2\frac{1}{3}$ inches.

Answers

Practice Exercises 3.4A:

1. $\dfrac{1}{4}$

2. $\dfrac{16}{33}$

3. $\dfrac{36}{5} = 7\dfrac{1}{5}$

4. 20

5. $\dfrac{15}{2} = 7\dfrac{1}{2}$

6. 8

7. $\dfrac{38}{5} = 7\dfrac{3}{5}$

8. $\dfrac{41}{15} = 2\dfrac{11}{15}$

Practice Exercises 3.4B:

1. $\dfrac{7}{3} = 2\dfrac{1}{3}$

2. $\dfrac{1}{10}$

3. $\dfrac{16}{19}$

4. $\dfrac{56}{51} = 1\dfrac{5}{51}$

5. $\dfrac{16}{45}$

6. $\dfrac{108}{17} = 6\dfrac{6}{17}$

7. $\dfrac{3}{4}$

8. 4

Practice Exercises 3.4C:

1. The grazing area is 3,600 m².
2. The company should budget $560 for billboard production.
3. 8 full objects can be assembled in 1 hour.
4. The length of the remaining piece is $2\dfrac{1}{4}$ meters.
5. 17 ft² = 17 square feet.
6. $10\dfrac{1}{2}$ in² = $10\dfrac{1}{2}$ inches squared.

Module 3: Fractions
Objective 3.5A – Add fractions

To add fractions with the same denominator, add the numerators and place the sum over the common denominator.

Example:

Add: $\dfrac{5}{12} + \dfrac{1}{12}$

Solution:

$$\dfrac{5}{12} + \dfrac{1}{12} = \dfrac{6}{12} = \dfrac{1}{2}$$

Note that the answer is reduced to simplest form. Always write your answer in simplest form.

To add fractions with different denominators, first rewrite the fractions as equivalent fractions with a common denominator. Then add the numerators and place the sum over the common denominator. The LCM of the denominators of the fractions is the **least common denominator (LCD)**.

Example:

Find the sum of $\dfrac{2}{3}$ and $\dfrac{1}{4}$.

Solution:

The LCM of 3 and 4 is 12. The LCD is 12.

$$\dfrac{2}{3} + \dfrac{1}{4} = \dfrac{8}{12} + \dfrac{3}{12}$$
$$= \dfrac{11}{12}$$

A mixed number is the sum of a whole number and a fraction. To add a whole number and a mixed number, add the whole numbers and add the fractional part. The sum of a whole number and a mixed number is a mixed number.

Example:

Add: $4\dfrac{5}{7} + 13$

Solution:

$$4\dfrac{5}{7} + 13 = 4 + 13 + \dfrac{5}{7} = 17 + \dfrac{5}{7} = 17\dfrac{5}{7}$$

To add two mixed numbers, add the whole numbers and add the fractional parts. Remember to reduce the sum to simplest form.

Example:

What is $5\dfrac{2}{3}$ added to $8\dfrac{5}{6}$?

Solution:

$$8\dfrac{5}{6} + 5\dfrac{2}{3} = 8\dfrac{5}{6} + 5\dfrac{4}{6}$$
$$= 13\dfrac{9}{6}$$
$$= 13\dfrac{3}{2}$$
$$= 13 + 1\dfrac{1}{2}$$
$$= 14\dfrac{1}{2}$$

Practice Exercises 3.5A:

1. Add: $\dfrac{7}{15} + \dfrac{2}{15}$

2. Add: $9 + \dfrac{7}{8}$

3. $\dfrac{2}{5} + \dfrac{1}{3} =$

4. $\dfrac{2}{3} + \dfrac{1}{2} =$

5. $\dfrac{4}{9} + \dfrac{1}{6} =$

6. Find the sum of 5 and $4\dfrac{1}{3}$.

7. Add $9\dfrac{5}{6}$ to $3\dfrac{1}{2}$.

8. Find the sum of $\dfrac{1}{4}, \dfrac{3}{5}$, and $\dfrac{5}{6}$.

9. Evaluate $x + y$ for $x = \dfrac{4}{15}$ and $y = \dfrac{3}{20}$.

10. Evaluate $a + b + c$ when $a = 2\dfrac{3}{5}$, $b = 5\dfrac{1}{3}$, and $c = 8$

11. Evaluate $x + y + z$ when $x = \dfrac{7}{8}$, $y = \dfrac{3}{4}$, and $z = 2\dfrac{5}{6}$.

Objective 3.5B – Subtract fractions

To subtract fractions with the same denominator, subtract the numerators and place the difference over the common denominator.

Example:

Subtract: $\dfrac{9}{10} - \dfrac{3}{10}$

Solution:

$$\dfrac{9}{10} - \dfrac{3}{10} = \dfrac{6}{10}$$
$$= \dfrac{3}{5}$$

To subtract fractions with different denominators, first rewrite the fractions as equivalent fractions with a common denominator. Then subtract the numerators and place the difference over the common denominator.

Example:

Subtract: $\dfrac{2}{3} - \dfrac{1}{6}$

Solution:

$$\dfrac{2}{3} - \dfrac{1}{6} = \dfrac{4}{6} - \dfrac{1}{6}$$
$$= \dfrac{3}{6}$$
$$= \dfrac{1}{2}$$

To subtract mixed numbers without borrowing, subtract the fractional parts and then subtract the whole numbers.

Example:

Find the difference of $19\frac{3}{4}$ and $8\frac{1}{12}$.

Solution:

$$19\frac{3}{4} - 8\frac{1}{12} = 19\frac{9}{12} - 8\frac{1}{12}$$
$$= 11\frac{8}{12} \quad \left(\frac{9}{12} - \frac{1}{12} = \frac{8}{12} \text{ and } 19 - 8 = 11\right)$$
$$= 11\frac{2}{3}$$

Subtraction of mixed numbers sometimes involves borrowing.

Example:

Subtract: $8\frac{3}{5} - 2\frac{3}{4}$.

Solution:

$$8\frac{3}{5} - 2\frac{3}{4} = 8\frac{12}{20} - 2\frac{15}{20}$$
$$= \left(7 + 1 + \frac{12}{20}\right) - 2\frac{15}{20} \quad \text{(Borrow 1 from 8.)}$$
$$= 7\frac{32}{20} - 2\frac{15}{20} \quad \text{(Add the 1 to } \frac{12}{20}. \text{ Write } 1\frac{12}{20} \text{ as } \frac{32}{20}.)$$
$$= 5\frac{17}{20}$$

Practice Exercises 3.5B:

1. Subtract: $\dfrac{7}{16} - \dfrac{5}{16}$

2. Subtract: $\dfrac{5}{6} - \dfrac{3}{10}$

3. Find $20\dfrac{4}{9}$ decreased by $3\dfrac{1}{9}$.

4. What is $8\dfrac{1}{2}$ minus $4\dfrac{2}{3}$?

5. Evaluate $x - y$ for $x = 18$ and $y = 3\dfrac{2}{7}$.

6. What is $\dfrac{5}{9}$ less $\dfrac{2}{27}$?

7. Find $4\dfrac{2}{3}$ less than $8\dfrac{5}{6}$.

8. Evaluate $a - b$ for $a = 7$ and $b = \dfrac{3}{7}$.

9. $8 - 1\dfrac{2}{3} =$

Objective 3.5C – Solve application problems

Some application problems can be solved with fraction addition and subtraction.

Examples:

1. If a $3\frac{2}{3}$-ft length is cut from a 10-foot board, how much of the board remains?

   Solution:

   This problem requires subtraction.

   $$10 - 3\frac{2}{3} = (9+1) - 3\frac{2}{3}$$
   $$= 9\frac{3}{3} - 3\frac{2}{3}$$
   $$= 6\frac{1}{3}$$

   $6\frac{1}{3}$ ft of the board remains.

2. Suppose that you wish to lose 12 lb over 3 months. If in the first month you lose $4\frac{2}{5}$ lb and in the second month you lose $3\frac{2}{3}$ lb, how much is left to lose during the third month?

   Solution:

   This problem requires that we first add $4\frac{2}{5}$ and $3\frac{2}{3}$, and it then requires that we subtract the sum from 12.

$$12 - \left(4\frac{2}{5} + 3\frac{2}{3}\right) = 12 - \left(4\frac{6}{15} + 3\frac{10}{15}\right)$$

$$= 12 - 8\frac{1}{15}$$

$$= 11\frac{15}{15} - 8\frac{1}{15}$$

$$= 3\frac{14}{15}$$

You must lose $3\frac{14}{15}$ lb during the third month.

Practice Exercises 3.5C:

1. A vat initially contains $28\frac{2}{3}$ lb of a gel. If $8\frac{5}{6}$ lb of the gel are removed, what weight of gel remains in the vat?

2. Suppose that you are paid \$12/hour. What is your total two-week pay if you work $23\frac{1}{3}$ hours the first week but only $2\frac{1}{2}$ hours the second week?

3. Lengths of $\frac{5}{9}$ ft and $9\frac{2}{3}$ ft are cut from a board of length 12 ft. What length of board remains?

4. A person can walk $3\frac{1}{4}$ miles in one hour. How many miles can the person walk in $1\frac{3}{4}$ hours?

Answers

Practice Exercises 3.5A

1. $\dfrac{3}{5}$

2. $9\dfrac{7}{8}$

3. $9\dfrac{1}{3}$

4. $13\dfrac{1}{3}$

5. $\dfrac{101}{60} = 1\dfrac{41}{60}$

6. $\dfrac{5}{12}$

7. $\dfrac{239}{15} = 15\dfrac{14}{15}$

8. $\dfrac{107}{24} = 4\dfrac{11}{24}$

Practice Exercises 3.5B:

1. $\dfrac{1}{8}$

2. $\dfrac{8}{15}$

3. $17\dfrac{1}{3}$

4. $3\dfrac{5}{6}$

5. $14\dfrac{5}{7}$

6. $\dfrac{13}{27}$

7. $\dfrac{25}{6} = 4\dfrac{1}{6}$

8. $\dfrac{46}{7} = 6\dfrac{4}{7}$

Practice Exercises 3.5C:

1. $19\dfrac{5}{6}$ lb of gel remains in the vat.

2. $310

3. $1\dfrac{7}{9}$ ft of board remains

4. $\dfrac{91}{16}$ miles $= 5\dfrac{11}{16}$ miles

Module 3: Fractions

Objective 3.6A – Multiply and divide positive and negative fractions

The sign rules for multiplying positive and negative fractions are the same rules used to multiply integers.

The product of two numbers with the same sign is positive.

The product of two numbers with different signs is negative.

Examples:

1. Find the product of $\dfrac{7}{9}$ and $-\dfrac{3}{5}$.

    Solution: $\left(\dfrac{7}{9}\right)\left(-\dfrac{3}{5}\right) = -\left(\dfrac{7}{9} \cdot \dfrac{3}{5}\right)$

    $= -\dfrac{7 \cdot 3}{3 \cdot 3 \cdot 5}$

    $= -\dfrac{7}{15}$

2. Evaluate xy for $x = -1\dfrac{1}{5}$ and $y = -3\dfrac{1}{3}$.

    Solution: xy

    $\left(-1\dfrac{1}{5}\right)\left(-3\dfrac{1}{3}\right) = \left(-\dfrac{6}{5}\right)\left(-\dfrac{10}{3}\right)$

    $= \dfrac{2 \cdot 3 \cdot 2 \cdot 5}{5 \cdot 3}$

    $= \dfrac{4}{1} = 4$

The sign rules for dividing positive and negative fractions are the same rules used to divide integers.

The quotient of two numbers with the same sign is positive.

The quotient of two numbers with different signs is negative.

Examples:

1. Find the quotient of -8 and $\frac{3}{4}$.

    Solution: $\quad -8 \div \frac{3}{4} = -\frac{8}{1} \cdot \frac{4}{3}$

    $\qquad\qquad\qquad = -\dfrac{2 \cdot 2 \cdot 2 \cdot 2 \cdot 2}{3}$

    $\qquad\qquad\qquad = -\dfrac{32}{3} = -10\dfrac{2}{3}$

2. Divide: $\quad -5\frac{3}{7} \div \left(-\frac{7}{10} \right)$

    Solution: $\quad -5\dfrac{3}{7} \div \left(-\dfrac{7}{10} \right) = -\dfrac{38}{7} \cdot \left(-\dfrac{7}{10} \right)$

    $\qquad\qquad\qquad\qquad = \dfrac{2 \cdot 19 \cdot 7}{7 \cdot 2 \cdot 5}$

    $\qquad\qquad\qquad\qquad = \dfrac{19}{5} = 3\dfrac{4}{5}$

Practice Exercises 3.6A:

1. Find the product of $-\dfrac{2}{3}$ and $\dfrac{6}{7}$.

2. Multiply: $\quad -1\dfrac{3}{5} \cdot -\dfrac{10}{13}$

3. What is $-2\dfrac{2}{9}$ times $-1\dfrac{4}{5}$?

4. Find the quotient of $-\dfrac{2}{3}$ and $\dfrac{6}{7}$.

5. Divide: $-4 \div \left(-\dfrac{6}{11}\right)$

6. Divide: $-7\dfrac{1}{8} \div \left(-3\dfrac{1}{2}\right)$

7. Evaluate $x \div y$ for $x = 9\dfrac{1}{2}$ and $y = -3$.

8. Evaluate xy for $x = -5\dfrac{2}{3}$ and $y = -\dfrac{9}{34}$.

9. Evaluate abc for $a = 2\dfrac{1}{5}$, $b = -\dfrac{3}{22}$, and $c = \dfrac{2}{9}$.

Objective 3.6B – Add and subtract positive and negative fractions

To add a fraction with a negative sign, rewrite the fraction with the negative sign in the numerator. Then add the numerators and place the sum over the common denominator.

Example: Add $\dfrac{1}{3}$ to $-\dfrac{5}{6}$.

Solution:

$$-\frac{5}{6} + \frac{1}{3} = \frac{-5}{6} + \frac{1}{3}$$

$$= \frac{-5}{6} + \frac{2}{6}$$

$$= \frac{-3}{6}$$

$$= \frac{-1}{2}$$

$$= -\frac{1}{2}$$

Although the answer could have been left as $\dfrac{-1}{2}$, we will write all negative fractions with the negative sign in front of the fraction.

To subtract fractions with negative signs, first rewrite the fractions with the negative signs in the numerators.

Example: Subtract $\dfrac{5}{12}$ from $-\dfrac{5}{6}$.

Solution:

$$-\frac{5}{6} - \frac{5}{12} = \frac{-5}{6} - \frac{5}{12}$$

$$= \frac{-10}{12} - \frac{5}{12}$$

$$= \frac{-10 - 5}{12}$$

$$= \frac{-15}{12}$$

$$= -\frac{5}{4} = -1\frac{1}{4}$$

Practice Exercises 3.6B:

1. Add: $-\dfrac{5}{8}+\dfrac{1}{3}$

2. Add: $-\dfrac{1}{3}+\dfrac{5}{12}$

3. Find the sum of $-\dfrac{1}{3},\dfrac{1}{2},$ and $-\dfrac{2}{5}$.

4. Subtract: $-\dfrac{5}{6}-\dfrac{2}{3}$

5. Subtract: $-\dfrac{7}{12}-\left(-\dfrac{5}{6}\right)$

6. Evaluate $a-b$ for $a=\dfrac{7}{10}$ and $b=-\dfrac{5}{8}$.

7. Evaluate $x+y$ for $x = -2\frac{1}{3}$ and $y = -\frac{7}{9}$.

8. Evaluate $a+b-c$ for $a = -4\frac{2}{5}$, $b = \frac{1}{10}$, and $c = -1\frac{2}{3}$.

Answers

Practice Exercises 3.6A

1. $-\dfrac{4}{7}$

2. $\dfrac{16}{13} = 1\dfrac{3}{13}$

3. 4

4. $-\dfrac{7}{9}$

5. $\dfrac{22}{3} = 7\dfrac{1}{3}$

6. $\dfrac{57}{28} = 2\dfrac{1}{28}$

7. $\dfrac{-19}{6} = -3\dfrac{1}{6}$

8. $\dfrac{3}{2} = 1\dfrac{1}{2}$

9. $-\dfrac{1}{15}$

Practice Exercises 3.6B:

1. $-\dfrac{7}{24}$

2. $\dfrac{1}{12}$

3. $-\dfrac{7}{30}$

4. $-\dfrac{3}{2} = -1\dfrac{1}{2}$

5. $\dfrac{1}{4}$

6. $\dfrac{53}{40} = 1\dfrac{13}{40}$

7. $-\dfrac{28}{9} = -3\dfrac{1}{9}$

8. $-\dfrac{79}{30} = -2\dfrac{19}{30}$

Module 3: Fractions

Objective 3.7A – Use the Order of Operations Agreement to simplify expressions

Recall that repeated multiplication of the same number is written using an exponent. In the **exponential expression** $\left(\dfrac{3}{4}\right)^2$, the exponent 2 indicates the number of times to multiply $\dfrac{3}{4}$.

$$\left(\frac{3}{4}\right)^2 = \frac{3}{4} \cdot \frac{3}{4} = \frac{9}{12}$$

The Order of Operations Agreement

Step 1: Perform operations inside grouping symbols such as parentheses and fraction bars.

Step 2: Simplify exponential expressions.

Step 3: Do multiplication and division as they occur from left to right.

Step 4: Do addition and subtraction as they occur from left to right.

One or more of the steps listed above may not be needed to evaluate an expression. In that case, proceed to the next step in the Order of Operations Agreement.

Examples:

1. Simplify: $\left(-\dfrac{3}{5}\right)^2 \div \left(-\dfrac{1}{3} - \dfrac{4}{15}\right)$

   Solution:

$$\left(-\frac{3}{5}\right)^2 \div \left(-\frac{1}{3} - \frac{4}{15}\right) = \frac{9}{25} \div \left(-\frac{1}{3} - \frac{4}{15}\right)$$

$$= \frac{9}{25} \div \left(\frac{-5}{15} - \frac{4}{15}\right)$$

$$= \frac{9}{25} \div \left(-\frac{3}{5}\right)$$

$$= \frac{9}{25} \cdot \left(-\frac{5}{3}\right)$$

$$= -\frac{3}{5}$$

2. Simplify: $-\left(\dfrac{3}{5}\right)^2 \div \left(-\dfrac{1}{3}\right) \cdot \left(-\dfrac{4}{15}\right)$

Solution:

$$-\left(\frac{3}{5}\right)^2 \div \left(-\frac{1}{3}\right) \cdot \left(-\frac{4}{15}\right) = -\frac{9}{25} \div \left(-\frac{1}{3}\right) \cdot \left(-\frac{4}{15}\right)$$

$$= -\frac{9}{25} \cdot \left(-\frac{3}{1}\right) \cdot \left(-\frac{4}{15}\right)$$

$$= -\frac{36}{125}$$

Practice Exercises 3.7A:

1. Simplify: $\left(-\dfrac{2}{3}\right)^3 - \dfrac{1-3}{2-8} + \dfrac{2}{5}$

2. Simplify: $-\dfrac{3}{4} \div \dfrac{1}{2} \cdot \dfrac{8}{21}$

3. Simplify: $\left(\dfrac{2}{5} - \dfrac{1}{3}\right)^2 \div \dfrac{2}{5} - \left(-\dfrac{2}{3}\right)$

4. Simplify: $\dfrac{1}{8} \div \left(\dfrac{1}{2}\right)^2 - \dfrac{1}{2}$

5. Simplify: $\left(\dfrac{1}{2}\right)^3 - \left(\dfrac{2}{5}\right)\left(\dfrac{15}{22}\right) + \dfrac{1}{4}$

6. Evaluate $x^2 y + z$ for $x = \dfrac{4}{7}$, $y = \dfrac{3}{4}$, and $z = \dfrac{11}{14}$

Objective 3.7B – Simplify complex fractions

A **complex fraction** is a fraction whose numerator or denominator contains one or more fractions.

Here are a few examples of complex fractions:

Main fraction bar → $\dfrac{\dfrac{1}{3}}{\dfrac{2}{5}}$ $\qquad \dfrac{4-\dfrac{3}{5}}{8} \qquad \dfrac{1\dfrac{1}{9}\cdot\dfrac{3}{5}}{-\dfrac{2}{3}-\left(-\dfrac{1}{2}\right)}$

To simplify a complex fraction, first simplify the expression above the main fraction bar and the expression below the main fraction bar; the result is one number in the numerator and one number in the denominator. Then rewrite the complex fraction as a division problem by reading the main fraction bar as "divided by."

Examples:

1. Simplify $\dfrac{\dfrac{1}{3}}{\dfrac{2}{5}}$.

   Solution: The expressions above and below the main fraction bar are already simplified. By reading the main fraction bar as "divided by" we have

$$\dfrac{\dfrac{1}{3}}{\dfrac{2}{5}} = \dfrac{1}{3} \div \dfrac{2}{5}$$

$$= \dfrac{1}{3} \cdot \dfrac{5}{2}$$

$$= \dfrac{5}{6}$$

2. Simplify: $\dfrac{4-\dfrac{3}{5}}{8}$

Solution:

$$\frac{4-\dfrac{3}{5}}{8} = \frac{\dfrac{20}{5}-\dfrac{3}{5}}{8}$$

$$= \frac{\dfrac{17}{5}}{8}$$

$$= \frac{17}{5} \div 8$$

$$= \frac{17}{5} \cdot \frac{1}{8}$$

$$= \frac{17}{40}$$

3. Simplify: $\dfrac{1\dfrac{1}{9} \cdot \dfrac{3}{5}}{-\dfrac{2}{3}-\left(-\dfrac{1}{2}\right)}$

Solution:

$$\frac{1\dfrac{1}{9} \cdot \dfrac{3}{5}}{-\dfrac{2}{3}-\left(-\dfrac{1}{2}\right)} = \frac{\dfrac{10}{9} \cdot \dfrac{3}{5}}{\dfrac{-2}{3}+\dfrac{1}{2}}$$

$$= \frac{\dfrac{2}{3}}{\dfrac{-4}{6}+\dfrac{3}{6}}$$

$$= \frac{\dfrac{2}{3}}{\dfrac{-1}{6}}$$

$$= \frac{2}{3} \cdot \left(-\frac{6}{1}\right)$$

$$= -4$$

Practice Exercises 3.7B:

1. Simplify: $\dfrac{\dfrac{3}{4}}{\dfrac{7}{8}}$

2. Simplify: $\dfrac{-\dfrac{4}{5}-\dfrac{2}{3}}{3+\dfrac{1}{2}}$

3. Simplify: $\dfrac{8-1\dfrac{1}{3}}{-2-4\dfrac{1}{2}}$

4. Simplify $\dfrac{xy}{z}$ for $x=3$, $y=-2\dfrac{1}{5}$, and $z=-\dfrac{3}{4}$.

5. Simplify $\dfrac{x+y}{z}$ for $x = \dfrac{3}{8}$, $y = \dfrac{2}{3}$, and $z = \dfrac{1}{12}$.

6. Simplify $\dfrac{a}{b-c}$ for $a = -\dfrac{5}{8}$, $b = \dfrac{1}{2}$, and $c = \dfrac{2}{5}$.

Answers

Practice Exercises 3.7A

1. $-\dfrac{31}{135}$

2. $-\dfrac{4}{7}$

3. $\dfrac{61}{90}$

4. 0

5. $\dfrac{9}{88}$

6. $\dfrac{101}{98} = 1\dfrac{3}{98}$

Practice Exercises 3.7B

1. $\dfrac{6}{7}$

2. $-\dfrac{44}{105}$

3. $-\dfrac{40}{39}$

4. $\dfrac{44}{5} = 8\dfrac{4}{5}$

5. $\dfrac{25}{2} = 12\dfrac{1}{2}$

6. $-\dfrac{25}{4} = -6\dfrac{1}{4}$

Module 4: Decimals and Percents
Objective 4.1A – Write decimals in standard form and in words

The number 72.85 is in **decimal notation**. In decimal notation, the part of the number that appears to the left of the decimal point is the **whole-number part**. Here, 72 is the whole-number part. The part of the number that appears to the right of the decimal point is the **decimal part**. Here, 85 is the decimal part. The **decimal point** separates the whole-number part from the decimal part. The decimal part of the number represents a number less than 1.

The place-value chart is extended to the right to show the place values of digits to the right of the decimal point.

Thousands	Hundreds	Tens	Ones		Tenths	Hundredths	Thousandths	Ten-thousandths	Hundred-thousandths
8	2	9	7	.	1	4	5	6	3

The digit 4 in the number 8297.14563 is in the hundredths place. The digit 6 is in the ten-thousandths place.

To write a decimal in words, write the decimal part of the number as though it were a whole number, and then name the place value of the last digit. The decimal point is read as "and." The number in the table above is eight thousand two hundred ninety-seven and fourteen thousand five hundred sixty-three hundred-thousandths.

Example:

   Write 32.8025 in words.

   Solution:

   Thirty-two and eight thousand twenty-five ten-thousandths

To write a decimal in standard form when it is written in words, write the whole-number part, replace the word and with a decimal point, and write the decimal part so that the last digit is in the given place-value position. When writing a decimal in standard form, you may need to insert zeros after the decimal point so that the last digit is in the given place-value position.

Example:

Write three hundred four and nineteen millionths in standard form.

Solution: 304.000019

Practice Exercises 4.1A:

1. Write the fraction as a decimal: $\dfrac{7}{100}$

2. Write the fraction as a decimal: $\dfrac{353}{1,000}$

3. Write the fraction as a decimal: $\dfrac{9}{10}$

4. Write the decimal as a fraction: 0.501

5. Write the decimal as a fraction: 0.07

6. Write the decimal as a fraction: 0.8411

7. Write the number in words: 0.39

8. Write the number in words: 2.007

9. Write the number in words: 26.379

10. Write the number in words: 514.3118

11. Write the number in standard form: eight hundred thirty-four thousandths

12. Write the number in standard form: six and one hundred one ten-thousandths

13. Write the number in standard form: twenty-five and seven thousand two hundred ninety-three hundred-thousandths

Objective 4.1B – Round a decimal to a given place value

In general, rounding decimals is similar to rounding whole numbers except that the digits to the right of the given place value are dropped instead of being replaced by zeros.

If the digit to the right of the given place value is less than 5, that digit and all digits to the right are dropped.

Example:

Round 44.38128 to the nearest thousandth.

Solution:

The given place value is the thousandths, and the digit 1 is in the thousandths place. The digit to the right, 2, is less than 5. The 2 and 8 are dropped. 44.38128 rounded to the nearest thousandth is 44.381.

If the digit to the right of the given place value is greater than or equal to 5, increase the digit in the given place value by 1, and drop all digits to its right.

Example:

Round 44.38128 to the nearest ten-thousandth.

Solution:

The given place value is the ten-thousandths, and the digit 2 is in the ten-thousandths place. The digit to the right, 8, is greater than 5. The 8 is drooped. 44.38128 rounded to the nearest ten-thousandth is 44.4813.

Practice Exercises 4.1B:

1. Round 0.064 to the tenths place.

2. Round 26.349 to the tenths place.

3. Round 65.34498 to the hundredths place.

4. Round 517.677 to the hundredths place.

5. Round 51.00439 to the thousandths place.

6. Round 4.37628 to the ten-thousandths place.

7. Round 0.009123 to the hundred-thousandths place.

8. Round 1.49256 to the nearest whole number.

9. Round 3.60021 to the nearest whole number.

10. Round 70.50648 to the nearest whole number.

Objective 4.1C – Compare decimals

There is a relationship between numbers written in decimal notation and fractions.

Examples:

$$0.59 = \frac{59}{100} \qquad\qquad 0.00127 = \frac{127}{100{,}000}$$

This relationship can be used to compare decimals. To compare decimals:

- Write the numbers as fractions.
- Write the fractions with a common denominator.
- Compare the fractions.

Example:

Place the correct symbol, < or >, between the numbers.

$$0.0257 \qquad\qquad 0.027$$

Solution:

$$0.0257 = \frac{257}{10{,}000} \qquad\qquad 0.027 = \frac{27}{1{,}000}$$

$$\frac{257}{10{,}000} \qquad\qquad \frac{27}{1{,}000} = \frac{270}{10{,}000}$$

$$\frac{257}{10{,}000} \quad < \quad \frac{270}{10{,}000}$$

$$0.0257 \quad < \quad 0.027$$

Practice Exercises 4.1C:

1. Place the correct symbol, < or >, between the numbers.

   0.23   0.3

2. Place the correct symbol, < or >, between the numbers.

   4.54   4.45

3. Place the correct symbol, < or >, between the numbers.

   7.10   7.01

4. Place the correct symbol, < or >, between the numbers.

   9.143   9.134

5. Place the correct symbol, < or >, between the numbers.

   0.091   0.101

6. Place the correct symbol, < or >, between the numbers.

   0.25   0.256

7. Place the correct symbol, < or >, between the numbers.

    0.63   0.063

8. Place the correct symbol, < or >, between the numbers.

    0.3   1.003

9. Place the correct symbol, < or >, between the numbers.

    0.7   0.079

10. Place the correct symbol, < or >, between the numbers.

    2.907   2.097

11. Write the given numbers in order from smallest to largest.
    0.0037, 0.037, 0.00037, 0.37

12. Write the given numbers in order from smallest to largest.
    0.11, 0.0001, 0.012, 0.21

Answers

Practice Exericses 4.1A:

1. 0.07
2. 0.353
3. 0.9
4. $\dfrac{501}{1,000}$
5. $\dfrac{7}{100}$
6. $\dfrac{8,411}{10,000}$

7. Thirty-nine hundredths
8. Two and seven thousandths
9. Twenty-six and three hundred seventy-nine thousandths
10. Five hundred fourteen and three thousand one hundred eighteen ten-thousandths
11. 0.834
12. 6.0101
13. 25.07293

Practice Exericses 4.1B:

1. 0.1
2. 26.3
3. 65.34
4. 517.68
5. 51.004
6. 4.3763
7. 0.00912
8. 1
9. 4
10. 71

Practice Exericses 4.1C:

1. 0.23 < 0.3
2. 4.54 > 4.45
3. 7.10 > 7.01
4. 9.143 > 9.134
5. 0.091 < 0.101
6. 0.63 > 0.063
7. 0.3 < 1.003
8. 0.7 > 0.079
9. 2.907 > 2.097
10. 0.00037, 0.0037, 0.037, 0.37
11. 0.00037, 0.0037, 0.037, 0.37
12. 0.0001, 0.012, 0.11, 0.21

Module 4: Decimals and Percents

Objective 4.2A – Add and subtract decimals

To add decimals, write the numbers so that the decimal points are on a vertical line. Add as you would with whole numbers. Then write the decimal point in the sum directly below the decimal points in the addends.

Example:

Find the sum of 8.238, 19, and 7.8974.

Solution:

$$
\begin{array}{r}
\overset{2}{}\,\overset{1}{}\,\overset{1}{3}\,\overset{1}{} \\
8\,.\,2\,3\,8 \\
1\,9\,. \\
+\quad 7\,.\,8\,9\,7\,4 \\
\hline
3\,5\,.\,1\,3\,5\,4
\end{array}
$$

The sum is 35.1354.

To subtract decimals, write the numbers so that the decimal points are on a vertical line. Subtract as you would with whole numbers. Then write the decimal point in the difference directly below the decimal point in the subtrahend. Insert zeros in the minuend or subtrahend, if necessary, so that each has the same number of decimal places.

Example:

Subtract: $8.35 - 2.9187$

Solution:

$$
\begin{array}{r}
\overset{7}{\cancel{8}}\,.\,\overset{13}{\cancel{3}}\,\overset{4}{\cancel{5}}\,\overset{9}{\cancel{0}}\,\overset{10}{\cancel{0}} \\
-\;2\,.\,9\,1\,8\,7 \\
\hline
5\,.\,4\,3\,1\,3
\end{array}
$$

The difference is 5.4313.

The sign rules for adding and subtracting decimals are the same rules used to add and subtract integers.

Practice Exercises 4.2A:

1. Add: $25.25 + 7.4418 + 18.5$

2. Subtract: $46.287 - 13.91$

3. Add: $6.841 + 54 + 59.3254$

4. Subtract: $145.03 - 8.2174$

5. Subtract: $650 - 56.413$

6. Add: $75.514 + 0.199 + 29 + 8.356$

7. Subtract: $9.08 - 6.324$

8. Evaluate $x + y + z$ for $x = 2.7156$, $y = 45.08$, and $z = 6.0406$.

9. Evaluate $x - y$ for $x = 0.847$ and $y = 0.25$.

Objective 4.2B – Solve application problems and use formulas

Examples:

Home Runs Hit for Every 100 At-Bats	
Harmon Killebrew	7.03
Ralph Kiner	7.09
Babe Ruth	8.05
Ted Williams	6.76

For Problems 1 and 2, use the table above.

1. According to the table, who had more home runs for every 100 at-bats, Harmon Killebrew or Ralph Kiner?

    Solution:

    Since $7.09 > 7.03$, Ralph Kiner had more home runs for every 100 at-bats than did Harmon Killebrew.

2. How many more home runs per 100 at-bats did Babe Ruth have than Ted Williams?

    Solution:

    Since $8.05 - 6.76 = 1.29$, Babe Ruth had 1.29 home runs per 100 at-bats more than did Ted Williams.

3. Use the formula $P = R - C$, where P is the profit for selling a product, R is the revenue, and C is the cost of producing the product, to find the profit for a product for which production costs are \$78,453.94 and revenue is \$100,000.

    Solution:

    $$P = R - C$$
    $$= 100,000 - 78,453.94$$
    $$= 21,546.06$$

    The profit is \$21,546.06.

Practice Exercises 4.2B:

1. An athlete bicycles 17.2 miles on Monday, 6.7 miles on Tuesday, and 12.4 miles on Wednesday. What was the total distance traveled for three days?

2. The odometer of your car reads 2,456.2 miles. you drive 65.4 miles on Friday, 56.9 miles on Saturday, and 73.3 miles on Sunday. Find your odometer reading at the end of the three days.

3. You buy a toaster for $16.83. How much change do you receive from a $20.00 bill?

4. A competitive swimmer beat the team's record time of 56.27 seconds in the 100-meter freestyle competition by 0.89 seconds. What is the new record time?

5. Use the formula $M = S - C$, where M is the markup on a consumer product, S is the selling price, and C is the cost of the product to the business, to find the markup on a product that costs a business $2,837.27 and has a selling price of $3,899.99.

6. Find the equity on a home that is valued at $327,000 when the homeowner has $186,249.20 in loans on the property. Use the formula $E = V - L$, where E is the equity, V is the is the value of the home, and L is the loan amount on the property.

Answers

Practice Exercises 4.2A:

1. 51.1918
2. 32.377
3. 120.1664
4. 136.8126
5. 593.587
6. 113.069
7. 2.756
8. 53.8362
9. 0.597

Practice Exercises 4.2B:

1. 36.3 miles
2. 2,651.8 miles
3. $3.17
4. 55.38 seconds
5. $1,062.72
6. $140,750.80

Module 4: Decimals and Percents

Objective 4.3A – Multiply decimals

To multiply decimals, multiply the numbers as you would whole numbers. Then write the decimal point in the product so that the number of decimal places in the product is the sum of the numbers of decimal places in the factors.

Example:

Multiply: $(0.361)(20.7)$

Solution:

$$
\begin{array}{r}
0.\ 3\ 6\ 1 \quad \text{3 decimal places} \\
\times \quad 2\ 0.\ 7 \quad \text{1 decimal place} \\
\hline
2\ 5\ 2\ 7 \\
7\ 2\ 2 \quad\quad\quad \\
\hline
7.\ 4\ 7\ 2\ 7 \quad \text{4 decimal places}
\end{array}
$$

The product is 7.4727.

To multiply a decimal by a power of 10 (10, 100, 1,000, . . .), move the decimal point to the right the same number of places as there are zeros in the power of 10.

Example:

Find the product of 43.21 and 10,000.

Solution:

Since 10,000 has 4 zeros, the decimal point in 43.21 is moved four places to the right.

$43.21 \times 10,000 = 432,100$

Note that if the power of 10 is written in exponential notation, the exponent indicates how many places to move the decimal point.

Example:

Find the product of 8.16 and 10^6.

Solution:

The exponent on 10^6 is 6. Move the decimal point in 8.16 6 digits to the right.

$$8.16 \times 10^6 = 8,160,000$$

The sign rules for multiplying decimals are the same rules used to multiply integers.

Practice Exercises 4.3A:

1. Multiply: $(6.4)(0.3)$

2. Find the product of 0.28 and 0.6.

3. Multiply: $(0.96)(3.7)$

4. Multiply: $(2.64)(0.03)$

5. Multiply: $(1.07)(0.066)$

6. Multiply: 5.92×0.8

7. Evaluate the expression $1,000c$, when $c = 6.8235$.

8. Evaluate the expression xy, when $x = 3.278$ and $y = 4.6$.

Objective 4.3B – Divide decimals

To divide decimals, move the decimal point in the divisor to the right so that the divisor is a whole number. Move the decimal point in the dividend the same number of places to the right. Place the decimal point in the quotient directly above the decimal point in the dividend. Then divide as you would with whole numbers.

Example:

Divide: $41.756 \div 5.72$

Solution:

Move the decimal point 2 places to the right in the divisor. Move the decimal point 2 places to the right in the dividend. Place the decimal point in the quotient. Then divide as shown below.

$$
\begin{array}{r}
7.3 \\
572{\overline{\smash{\big)}\,4{,}175.6}} \\
\underline{-4{,}004} \\
1{,}71\,6 \\
\underline{-1{,}71\,6} \\
0
\end{array}
$$

In division of decimals, rather than writing the quotient with a remainder, we usually round the quotient to a specified place value. The symbol $\approx$ is read "is approximately equal to"; it is used to indicate that the quotient is an approximate value after being rounded.

Example:

Divide and round to the nearest tenth: $0.47 \div 0.6$

Solution:

Moving the decimal points in the dividend and divisor one place to the right gives us the work below:

$$\begin{array}{r} 0.78 \\ 6\overline{)4.70} \\ -4\,2 \\ \hline 50 \\ -48 \\ \hline 2 \end{array}$$

To round the quotient to the nearest tenth, the division must be carried to the hundredths place. Therefore, a zero must be inserted in the dividend so that the quotient has a digit in the hundredths place.

$$0.47 \div 0.6 \approx 0.8$$

To divide a decimal by a power of 10 (10, 100, 1,000, 10,000, . . .), move the decimal point to the left the same number of places as there are zeros in the power of 10.

Example:

Find the quotient of 43.21 and 10,000.

Solution:

Since 10,000 has 4 zeros, the decimal point in 43.21 is moved four places to the left.

$$43.21 \div 10,000 = 0.004321$$

If the power of 10 is written in exponential notation, the exponent indicates how many places to move the decimal point.

Example:

Find the quotient of 8.16 and 10^6.

Solution:

The exponent on 10^6 is 6. Move the decimal point in 8.16 6 digits to the left.

$$8.16 \div 10^6 = 0.00000816$$

The sign rules for dividing decimals are the same rules used to divide integers.

Practice Exercises 4.3B:

1. Divide: $5.64 \div 6$

2. Divide: $40 \div 0.8$

3. Divide: $6.515 \div 5$

4. Divide: $0.1116 \div 0.012$

5. Divide and round to the nearest tenth: $73.85 \div 9.6$

6. Divide and round to the nearest tenth: $50 \div 0.8$

7. Divide and round to the nearest hundredth: $4.724 \div 17$

8. Divide and round to the nearest thousandth: $0.0717 \div 0.9$

9. Divide and round to the nearest whole number: $34.19 \div 36$

10. Evaluate $x \div y$, where $x = 2.898$ and $y = 0.92$.

Objective 4.3C – Solve application problems and use formulas

Example:

If you purchase 4 items that each cost $1.72 with a $10 bill, how much change should you expect?

Solution:

We should subtract the product of $1.72 and 4 from $10.

$1.72 \times 4 = 6.88$

$10 - 6.88 = 3.12$

You should expect $3.12 change.

Example:

Use the formula $d = rt$, where d is the distance, r is the rate, and t is the time, to compute the distance traveled by an object moving at a rate of 9.23 m/s for 8.2 s. Round to the nearest tenth.

Solution:

We replace r with 9.23 and t with 8.2 in the formula to solve.

$d = rt$
$d = (9.23)(8.2)$
$d = 75.686$
$d \approx 75.7$

The object traveled approximately 75.7 meters.

Practice Exercises 4.3C:

1. A shuttle bus transports students from a suburban college to work study jobs in the city and back again 5 times a day. If the distance between the college and the city is 11.3 miles, find the distance the shuttle bus travels in one day.

2. A car is bought for $3,600 down and payments of $141.50 each month for 24 months. Find the total cost of the car.

3. You pay $1,284.72 per year in life insurance premiums. You pay the premiums in 12 equal monthly payments. Find the amount of each monthly payment.

4. Gasoline tax is $0.19 per gallon. Find the number of gallons used during a month in which $158.84 was paid in taxes.

5. A tax of $1.39 is paid on each hair dryer sold by a store. This month the total tax paid on hair dryers was $31.97. How many hair dryers were sold?

6. A jogger ran 6.8 miles in 42.16 minutes. What was the jogger's average time per mile?

Answers

Practice Exercises 4.3A:

1. 1.92
2. 0.168
3. 3.552
4. 0.0792
5. 0.07062
6. 4.736
7. 6,823.5
8. 15.0788

Practice Exercises 4.3B:

1. 0.94
2. 50
3. 1.303
4. 9.3
5. 7.7
6. 0.9
7. 0.28
8. 0.080
9. 1
10. 3.15

Practice Exercises 4.3C:

1. 113 miles
2. $6,996
3. $107.06
4. 836 gallons
5. 23 hairdryers
6. 6.2 minutes

Module 4: Decimals and Percents

Objective 4.4A – Convert fractions to decimals

To convert a fraction to a decimal, divide the numerator by the denominator. If the decimal eventually has a remainder of 0, then the decimal is a **terminating decimal**. If the remainder is never 0, then the decimal is a **repeating decimal**. It is common practice to write a bar over the repeating digits of a decimal.

Examples:

1. Convert $\dfrac{3}{16}$ to a decimal.

    Solution:

$$
\begin{array}{r}
0.1875 \\
16\overline{)3.0000} \\
-16 \\
\hline
140 \\
-128 \\
\hline
120 \\
-112 \\
\hline
80 \\
-80 \\
\hline
0
\end{array}
$$

$\dfrac{3}{16} = 0.1875$

2. Convert $\dfrac{7}{12}$ to a decimal.

    Solution:

$$
\begin{array}{r}
0.5833 \\
12\overline{)7.0000} \\
-60 \\
\hline
100 \\
-96 \\
\hline
40 \\
-36 \\
\hline
40 \\
-36 \\
\hline
4
\end{array}
$$

$\dfrac{7}{12} = 0.58\overline{3}$

Practice Exercises 4.4A:

1. Convert $\dfrac{5}{6}$ to a decimal. Place a bar over any repeating digits.

2. Convert $\dfrac{7}{16}$ to a decimal. Place a bar over any repeating digits.

3. Convert $\dfrac{3}{37}$ to a decimal. Place a bar over any repeating digits.

4. Convert $\dfrac{7}{13}$ to a decimal. Place a bar over any repeating digits.

5. Convert $\dfrac{13}{7}$ to a decimal. Place a bar over any repeating digits.

6. Convert $3\frac{2}{5}$ to a decimal. Place a bar over any repeating digits.

7. Convert $7\frac{2}{3}$ to a decimal. Place a bar over any repeating digits.

8. Convert $2\frac{1}{8}$ to a decimal. Place a bar over any repeating digits.

Objective 4.4B – Convert decimals to fractions

To convert a decimal to a fraction, remove the decimal point and place the decimal part over a denominator equal to the place value of the last digit in the decimal. Then write the fraction in simplest form.

Examples:

1. Convert 0.368 to a fraction.

   Solution:

   $$0.368 = \frac{368}{1,000} = \frac{\overset{1}{\cancel{2}} \cdot \overset{1}{\cancel{2}} \cdot \overset{1}{\cancel{2}} \cdot 2 \cdot 23}{\underset{1}{\cancel{2}} \cdot \underset{1}{\cancel{2}} \cdot \underset{1}{\cancel{2}} \cdot 5 \cdot 5 \cdot 5} = \frac{46}{125}$$

2. Convert 0.91 to a fraction.

   Solution:

   $$0.91 = \frac{7 \cdot 13}{100} = \frac{91}{100}$$

3. Convert 3.25 to a mixed number.

   Solution:

   $$3.25 = 3\frac{25}{100} = 3\frac{1}{4}$$

Practice Exercises 4.4B:

1. Convert 0.8 to a fraction.

2. Convert 0.72 to a fraction.

3. Convert 0.735 to a fraction.

4. Convert 2.25 to a mixed number.

5. Convert 0.096 to a fraction.

6. Convert 0.00032 to a fraction.

7. Convert 7.05 to a mixed number.

8. Convert $2.\overline{3}$ to a mixed number.

9. Convert 1.48 to a mixed number.

10. Convert 0.002 to a fraction.

Objective 4.4C – Compare a fraction and a decimal

One way to determine the order relation between a decimal and a fraction is to write the decimal as a fraction and then compare the fractions.

Example:

Place the correct symbol, < or >, between the two numbers.

$0.3 \quad \dfrac{1}{4}$

Solution:

$$0.3 \quad \dfrac{1}{4}$$

$$\dfrac{3}{10} \quad \dfrac{1}{4}$$

$$\dfrac{6}{20} \quad \dfrac{5}{20}$$

$$\dfrac{6}{20} > \dfrac{5}{20}$$

$$0.3 > \dfrac{1}{4}$$

Practice Exercises 4.4C:

1. Place the correct symbol, < or >, between the two numbers.

$$0.34 \qquad \dfrac{7}{20}$$

2. Place the correct symbol, < or >, between the two numbers.

$$\dfrac{11}{12} \qquad 0.85$$

3. Place the correct symbol, < or >, between the two numbers.

$$0.75 \qquad \dfrac{29}{40}$$

4. Place the correct symbol, < or >, between the two numbers.

$$\frac{3}{5} \qquad 0.65$$

5. Place the correct symbol, < or >, between the two numbers.

$$0.3 \qquad \frac{1}{3}$$

Objective 4.4D – Write ratios and rates

A **ratio** is the quotient or comparison of two quantities with the *same* unit.

A ratio can be written in three ways:

1. As a fraction
2. As two numbers separated by a colon
3. As two numbers separated by the word *to*

A ratio is in **simplest form** when the two number do not have a common factor. The units are not written in a ratio.

Example:

Write the comparison 9 miles to 12 miles as a ratio in simplest form using a fraction, a colon, and the word *to*.

Solution:

$$\frac{9 \text{ miles}}{12 \text{ miles}} = \frac{9}{12} = \frac{3}{4}$$

9 miles : 12 miles $= 9:12 = 3:4$

9 miles to 12 miles $= 9$ to $12 = 3$ to 4

A **rate** is the comparison of two quantities with *different* units. A rate is in **simplest form** when the numbers have no common factors. The units are written as part of the rate.

Many rates are written as **unit rates**. A unit rate is a rate in which the number in the denominator is 1.

Example:

Write "190 km in 2 h" as a unit rate.

Solution:

$$\frac{190 \text{ km}}{2 \text{ h}}$$
$190 \div 2 = 85$

The unit rate is 85 km/h.

Practice Exercises 4.4D:

1. Write the comparison 24 yards to 30 yards as a ratio in simplest form using a fraction, a colon, and the word *to*.

2. Write the comparison of 18 meters to 16 meters as a ratio in simplest form using a fraction, a colon, and the word *to*.

3. Write 30 miles in 8 hours as a unit rate.

4. Write $78 for 15 ounces as a unit rate.

5. A vehicle uses 15 gallons of gasoline to travel 537 miles. How many miles per gallon did the car get?

6. If Sally types 420 words in 6 minutes, how many words can she type in one minute?

Answers

Practice Exercises 4.4A:

1. $0.8\overline{3}$
2. 0.4375
3. $0.\overline{081}$
4. $0.\overline{538461}$
5. $1.\overline{857142}$
6. 3.4
7. $7.\overline{6}$
8. 2.125

Practice Exercises 4.4C:

1. $<$
2. $>$
3. $>$
4. $<$
5. $<$

Practice Exercises 4.4B:

1. $\dfrac{4}{5}$
2. $\dfrac{18}{25}$
3. $\dfrac{147}{200}$
4. $2\dfrac{1}{4}$
5. $\dfrac{12}{125}$
6. $\dfrac{1}{3,125}$
7. $7\dfrac{1}{20}$
8. $2\dfrac{1}{3}$
9. $1\dfrac{12}{25}$
10. $\dfrac{1}{500}$

Practice Exercises 4.4D:

1. $\dfrac{4}{5}$ $4:5$ 4 to 5
2. $\dfrac{9}{8}$ $9:8$ 9 to 8
3. 3.75 miles/hour
4. \$5.20/ounce
5. 35.8 miles/gallon
6. 70 words/minute

Module 4: Decimals and Percents

Objective 4.5A – Write a percent as a decimal or a fraction

Percent means "part of 100." The symbol % is the **percent sign**.

To write a percent as a decimal, remove the percent sign and multiply by 0.01.

To write a percent as a fraction, remove the percent sign and multiply by $\frac{1}{100}$.

Examples:

1. Write 135% as a decimal and as a fraction.

   Solution:

   $$135\% = 135 \times 0.01 = 1.35$$

   $$135\% = 135 \times \frac{1}{100} = \frac{135}{100} = \frac{27}{20}$$

2. Write $7\frac{8}{9}\%$ as a fraction.

   Solution:

   $$7\frac{8}{9}\% = 7\frac{8}{9} \times \frac{1}{100} = \frac{71}{9} \times \frac{1}{100} = \frac{71}{900}$$

Practice Exercises 4.5A:

1. Write 8.5% as a decimal and as a fraction.

2. Write 0.5% as a decimal and as a fraction.

3. Write 28% as a decimal and as a fraction.

4. Write 3.5% as a decimal.

5. Write $66\frac{2}{3}\%$ as a fraction.

6. Write $3\frac{1}{8}\%$ as a fraction.

7. Write 37% as a fraction.

8. Write 12% as a fraction.

Objective 4.5B – Write a decimal or a fraction as a percent

A decimal or a fraction can be written as a percent by multiplying by 100%.

When changing a fraction to a percent, if the fraction can be written as a terminating decimal, the percent is written in decimal form. If the decimal representation of the fraction is a repeating decimal, the answer is written with a fraction.

Examples:

1. Write 0.875 as a percent.

    Solution: $0.875 = 0.875 \times 100\% = 87.5\%$

2. Write $\dfrac{3}{16}$ as a percent.

    Solution: $\dfrac{3}{16} = \dfrac{3}{16} \times \dfrac{100\%}{1} = \dfrac{300\%}{16} = 18.75\%$

3. Write $\dfrac{1}{15}$ as a percent.

    Solution: $\dfrac{1}{15} = \dfrac{1}{15} \times \dfrac{100\%}{1} = \dfrac{100\%}{15} = 6\dfrac{2}{3}\%$

Practice Exercises 4.5B:

1. Write 0.094 as a percent.

2. Write $\dfrac{1}{8}$ as a percent.

3. Write $\dfrac{5}{12}$ as a percent.

4. Write 9.04 as a percent.

5. Write $\dfrac{21}{40}$ as a percent.

6. Write $\dfrac{9}{8}$ as a percent.

7. Write $\frac{4}{5}$ as a percent.

8. Write $\frac{7}{10}$ as a percent.

Answers

Practice Exercises 4.5A:

1. 0.085, $\dfrac{17}{200}$

2. 0.005, $\dfrac{1}{200}$

3. 0.28, $\dfrac{7}{25}$

4. 0.035

5. $\dfrac{2}{3}$

6. $\dfrac{1}{32}$

7. $\dfrac{37}{100}$

8. $\dfrac{3}{25}$

Practice Exercises 4.5B

1. 9.4%
2. 12.5%
3. $41\dfrac{2}{3}\%$
4. 904%
5. 52.5%
6. 112.5%
7. 80%
8. 70%

Module 4: Decimals and Percents

Objective 4.6A – Find the square root of a perfect square

Recall that the square of a number is equal to the number multiplied by itself. The square of an integer is called a **perfect square**. For example, 25 is a perfect square because 25 is the square of 5: $5^2 = 25$.

A **square root** of a positive number x is a number whose square is x. The symbol $\sqrt{}$ is used to indicate the positive square root of a number.

$$\sqrt{25} = 5 \quad \text{and} \quad -\sqrt{25} = -5$$

The square root symbol, $\sqrt{}$, is also called a **radical**. The number under the radical is called a **radicand**. In the radical expression $\sqrt{25}$, 25 is the radicand.

Examples:
1. Simplify $\sqrt{144}$.

    Solution: Since $12^2 = 144$, $\sqrt{144} = 12$.

2. Simplify $6 + 5\sqrt{16}$.

    Solution: $6 + 5\sqrt{16} = 6 + 5 \cdot 4$
    $$= 6 + 20$$
    $$= 26$$

3. Simplify $\sqrt{25} - 2\sqrt{4}$.

    Solution: $\sqrt{25} - 2\sqrt{4} = 5 - 2 \cdot 2$
    $$= 5 - 4$$
    $$= 1$$

4. Evaluate $3\sqrt{xy}$ for $x = 3$ and $y = 12$.

Solution: $3\sqrt{xy}$

$$3\sqrt{3\cdot 12} = 3\sqrt{36}$$
$$= 3\cdot 6$$
$$= 18$$

Practice Exercises 4.6A:
1. Simplify: $\sqrt{49}$

2. Simplify: $-\sqrt{16}$

3. Simplify: $-\sqrt{121}$

4. Simplify: $\sqrt{9+40}$

5. Simplify: $\sqrt{64} + \sqrt{25}$

6. Simplify: $-4\sqrt{81}$

7. Simplify: $\sqrt{\dfrac{4}{49}}$

8. Simplify: $\sqrt{\dfrac{1}{49}} - \sqrt{\dfrac{1}{81}}$

9. Simplify: $\sqrt{\dfrac{1}{16}} + \sqrt{\dfrac{1}{25}}$

10. Simplify: $\sqrt{25} + \sqrt{1}$

11. Evaluate $-5\sqrt{xy}$ for $x = 2$ and $y = 8$.

12. Evaluate $\sqrt{b^2 - 4ac}$ for $a = 2$, $b = 7$, and $c = -4$.

Answers

Practice Exercises 4.6A:

1. 7
2. −4
3. −11
4. 7
5. 13
6. −36
7. $\dfrac{2}{7}$
8. $\dfrac{2}{63}$
9. $\dfrac{9}{20}$
10. 6
11. −20
12. 9

5 Variable Expressions

Module 5: Variable Expressions

Objective 5.2A – Simplify a variable expression using the Properties of Addition

Like terms of a variable expression are terms with the same variable part. To simplify a variable expression, we use the Distributive Property to add the numerical coefficients of like variable terms. The variable part remains unchanged. This is called **combining like terms**.

The Distributive Property

If a, b, and c are real numbers, then $a(b+c) = ab + ac$ or $(b+c)a = ba + ca$.

The following Properties of Addition are used to simplify variable expressions.

The Associative Property of Addition

If a, b, and c are real numbers, then $(a+b)+c = a+(b+c)$.

The Commutative Property of Addition

If a and b are real numbers, then $a+b = b+a$.

The Addition Property of Zero

If a is a real number, then $a+0 = a$ and $0+a = a$.

The Inverse Property of Addition

If a is a real number, then $a+(-a) = 0$ and $(-a)+a = 0$.

Example:

Simplify $3x^2 - xy - 5y^2 + y^2 - 4x^2 + 2xy$.

Solution: Use the Commutative and Associative Properties of
              Addition to rearrange and group like terms.

$$3x^2 - xy - 5y^2 + y^2 - 4x^2 + 2xy$$
$$= \left(3x^2 - 4x^2\right) + \left(-xy + 2xy\right) + \left(-5y^2 + y^2\right)$$
$$= -x^2 + xy - 4y^2$$

Practice Exercises 5.2A

1. Simplify: $4x + 7x$

2. Simplify: $7a - 2a$

3. Simplify: $7ab - 5ab$

4. Simplify: $9xy - 11xy$

5. Simplify: $-8ab + 13ab$

6. Simplify: $-4mn + 4mn$

7. Simplify: $-\dfrac{1}{3}x - \dfrac{1}{4}x$

8. Simplify: $\dfrac{1}{4}x^2 - \dfrac{7}{12}x^2$

9. Simplify: $-\dfrac{2}{3}x^2 - \dfrac{3}{4}x^2$

10. Simplify: $-3x^2 - 8x^2 - 2x^2$

11. Simplify: $6x - 5x + 2y$

12. Simplify: $7y + 7x - 7y$

13. Simplify: $9y - 11x - 6y + 4y$

14. Simplify: $-4b + 3a - 9b + 11a$

Module 5: Variable Expressions
Objective 5.2B – Simplify a variable expression using the Properties of Multiplication

The following Properties of Multiplication are used to simplify variable expressions.

The Associative Property of Addition

If a, b, and c are real numbers, then $(ab)c = a(bc)$.

The Commutative Property of Addition

If a and b are real numbers, then $ab = ba$.

The Multiplication Property of One

If a is a real number, then $a \cdot 1 = a$ and $1 \cdot a = a$.

The Inverse Property of Addition

If a is a real number and a is not equal to zero, then $a \cdot \dfrac{1}{a} = 1$ and $\dfrac{1}{a} \cdot a = 1$.

$\dfrac{1}{a}$ is called the **reciprocal** or **multiplicative inverse** of a. The product of a number and its reciprocal is 1.

Examples:

1. Simplify: $\dfrac{7}{5}\left(\dfrac{5x}{7}\right)$

    Solution: $\dfrac{7}{5}\left(\dfrac{5x}{7}\right) = \dfrac{7}{5}\left(\dfrac{5}{7}x\right)$

    $\qquad\qquad = \left(\dfrac{7}{5} \cdot \dfrac{5}{7}\right)x$

    $\qquad\qquad = 1 \cdot x$

    $\qquad\qquad = x$

2. Simplify: $(3g)\left(-\dfrac{5}{6}\right)$

   Solution: $(3g)\left(-\dfrac{5}{6}\right) = 3\left(-\dfrac{5}{6}\right)g$

   $\qquad\qquad\qquad\quad = \left(-3 \cdot \dfrac{5}{6}\right)g$

   $\qquad\qquad\qquad\quad = -\dfrac{5}{2}g$

Practice Exercises 5.2B

   1. Simplify: $2\left(5x\right)$

   2. Simplify: $-3\left(2a\right)$

   3. Simplify: $-7\left(-8y\right)$

   4. Simplify: $-7\left(9x^2\right)$

   5. Simplify: $\dfrac{1}{4}\left(4x^2\right)$

6. Simplify: $\dfrac{1}{10}(10x)$

7. Simplify: $-\dfrac{1}{5}(-5a)$

8. Simplify: $(7x)\left(\dfrac{1}{7}\right)$

9. Simplify: $(-8y)\left(-\dfrac{1}{8}\right)$

10. Simplify: $\dfrac{1}{5}(15x)$

11. Simplify: $-\dfrac{1}{3}(12x)$

12. Simplify: $-\dfrac{7}{8}(32a^2)$

Module 5: Variable Expressions
Objective 5.2C – Simplify a variable expression using the Distributive Property

Recall that the Distributive Property states that if a, b, and c are real numbers, then $a(b+c) = ab + ac$. The Distributive Property is used to remove parentheses from a variable expression.

Examples:

1. Simplify: $-\left(4x^2 - 2x - 3\right)$

   Solution: $-\left(4x^2 - 2x - 3\right) = -1\left(4x^2 - 2x - 3\right)$
   $$= -1\left(4x^2\right) - (-1)(2x) - (-1)(3)$$
   $$= -4x^2 + 2x + 3$$

2. Simplify: $-4\left(-x - 2y + \dfrac{3}{2}z\right)$

   Solution: $-4\left(-x - 2y + \dfrac{3}{2}z\right) = -4(-x) - 4(-2y) - 4\left(\dfrac{3}{2}z\right)$
   $$= 4x + 8y - 6z$$

Practice Exercises 5.2C:

1. Simplify: $-(x + 5)$

2. Simplify: $4(2x - 3)$

3. Simplify: $-4(a+12)$

4. Simplify: $-4(3y-4)$

5. Simplify: $4(-5x^2-2)$

6. Simplify: $-7(3y^2-10)$

7. Simplify: $-5(x^2-y^2)$

8. Simplify: $-2\left(5a^2 - 8b^2\right)$

9. Simplify: $3\left(x^2 - 2x - 5\right)$

10. Simplify: $3\left(-a^2 - a - 4\right)$

11. Simplify: $-2\left(-3x^2 + 4x - 1\right)$

12. Simplify: $3\left(2x^2 - 3xy - 4y^2\right)$

13. Simplify: $-\left(9b^2 - 5b + 8\right)$

Module 5: Variable Expressions
Objective 5.2D – Simplify general variable expressions

When simplifying variable expressions, use the Distributive Property to remove parentheses and brackets used as grouping symbols.

Examples:

1. Simplify: $4(2-3y)-(3-2y)$

   Solution: $4(2-3y)-(3-2y) = 8-12y-3+2y$
   $$= 5-10y$$

2. Simplify: $3x-2\left[4x-3(-x-5)\right]$

   Solution: $3x-2\left[4x-3(-x-5)\right] = 3x-2\left[4x+3x+15\right]$
   $$= 3x-2\left[7x+15\right]$$
   $$= 3x-14x-30$$
   $$= -11x-30$$

Practice Exercises 5.2D

1. Simplify: $5x-3(2x+7)$

2. Simplify: $9-(10x-4)$

3. Simplify: $6 - (8 + 5y)$

4. Simplify: $3x - (10 - x)$

5. Simplify: $3(x - 3) - 2(x + 4)$

6. Simplify: $5(2y - 5) - 2(4 - y)$

7. Simplify: $3(a + b) - (a - 2b)$

8. Simplify: $5\left[x + 3(x + 6)\right]$

9. Simplify: $-4\left[x + 2(6 - x)\right]$

10. Simplify: $-5\left[2x - (4x + 1)\right]$

11. Simplify: $-6x + 4\left[x - 6(2 - x)\right]$

12. Simplify: $3a - 2\left[3b - (2b - a)\right] + 4b$

13. Simplify: $3x + 2(x - 2y) + 4(2x - 5y)$

Answers

Practice Exercises 5.2A

1. $11x$

2. $5a$

3. $2ab$

4. $-2xy$

5. $5ab$

6. 0

7. $-\dfrac{7}{12}x$

8. $-\dfrac{1}{3}x^2$

9. $-\dfrac{17}{12}x^2$

10. $-13x^2$

11. $x + 2y$

12. $7x$

13. $7y - 11x$

14. $-13b + 14a$

Practice Exercises 5.2C

1. $-x - 5$

2. $8x - 12$

3. $-4a - 48$

4. $-12y + 16$

5. $-20x^2 - 8$

6. $-21y^2 + 70$

7. $-5x^2 + 5y^2$

8. $-10a^2 + 16b^2$

9. $3x^2 - 6x - 15$

10. $-3a^2 - 3a - 12$

11. $6x^2 - 8x + 2$

12. $6x^2 - 9xy - 12y^2$

13. $-9b^2 + 5b - 8$

Practice Exercises 5.2B

1. $10x$

2. $-6a$

3. $56y$

4. $-63x^2$

5. x^2

6. x

7. a

8. x

9. y

10. $3x$

11. $-4x$

12. $-28a^2$

Practice Exercises 5.2D

1. $-x - 21$

2. $13 - 10x$

3. $-2 - 5y$

4. $4x - 10$

5. $x - 17$

6. $12y - 33$

7. $2a + 5b$

8. $20x + 90$

9. $4x - 48$

10. $10x + 5$

11. $22x - 48$

12. $a + 2b$

13. $13x - 24y$

Module 5: Variable Expressions

Objective 5.3A – Translate a verbal expression into a variable expression, given the variable

Some examples of verbal expressions for addition, along with their variable expressions, follow:

"9 added to x"	$x + 9$
"4 more than r"	$r + 4$
"the sum of 4 and d"	$4 + d$
"q increased by r"	$q + r$
"the total of 3 and g"	$3 + g$
"19 plus b"	$19 + b$

Some examples of verbal expressions for subtraction, along with their variable expressions, follow:

"y minus 9"	$y - 9$
"8 less than z"	$z - 8$
"8 less z"	$8 - z$
"23 subtracted from w"	$w - 23$
"9 decreased by h"	$9 - h$
"the difference between x and y"	$x - y$

Some examples of verbal expressions for multiplication, along with their variable expressions, follow:

"2 times r"	$2r$
"two-thirds of p"	$\dfrac{2}{3}p$
"the product of x and y"	xy
"u multiplied by 12"	$12u$
"twice n"	$2n$

Some examples of verbal expressions involving division, along with their variable expressions, follow:

"9 divided by x"	$\dfrac{9}{x}$
"the quotient of x and y"	$\dfrac{x}{y}$

"the ratio of 3 and m" $\dfrac{3}{m}$

Some examples of verbal expressions involving powers, along with their variable expressions, follow:

"the square of w" w^2
"the cube of n" n^3

Examples:

1. Translate "the difference of 5 times m and 3" into a variable expression.

   Solution: $5m - 3$

2. Translate "5 times the difference of m and 3" into a variable expression.

   Solution: $5(m - 3)$

3. Translate "12 decreased by the product of 9 and z" into a variable expression.

   Solution: $12 - 9z$

4. Translate "12 decreased by the sum of 9 and z" into a variable expression.

   Solution: $12 - (9 + z)$

Practice Exercises 5.3A

1. Translate "the sum of 6 and x" in to a varibale expression.

2. Translate "x less than 12" into a variable expression.

3. Translate "*a* decreased by 1" into a variable expression.

4. Translate "*z* divided by 8" into a variable expression.

5. Translate "10 more than the square of *x*" into a variable expression.

6. Translate "5 times the sum of *n* and 8" into a variable expression.

7. Translate "*x* increased by the product of 3 and *x*" into a variable expression.

8. Translate "the product of –2 and *t*" into a variable expression.

9. Translate "the product of 5 and the total of y and 6" into a variable expression.

10. Translate "12 more than one-fourth of the square of y" into a variable expression.

11. Translate "x increased by the quotient of x and 2" into a variable expression.

12. Translate "y decreased by the product of y and 2" into a variable expression.

Module 5: Variable Expressions

Objective 5.3B – Translate a verbal expression into a variable expression and then simplify

In most applications that involve translating phrases into variable expressions, the variable to be used is not given. To translate these phrases, a variable must be assigned to an unknown quantity before the variable expression can be written.

Examples:

1. Translate "three subtracted from one-half of the sum of 5 and the square of a number" into a variable expression. Then simplify.

    Solution:

    the square of a number: n^2
    the sum of 5 and the square of a number: $5 + n^2$
    one-half of the sum of 5 and the square of a number:

    $$\frac{1}{2}\left(5 + n^2\right)$$

    $$\frac{1}{2}\left(5 + n^2\right) - 3 = \frac{5}{2} + \frac{1}{2}n^2 - 3 = -\frac{1}{2} + \frac{1}{2}n^2$$

2. Translate "twice the difference of one-third of a number and the cube of 2, decreased by 4" into a variable expression. Then simplify.

    Solution:

    one-third of a number: $\frac{1}{3}n$

    the difference of one-third of a number and the cube of 2:

    $$\frac{1}{3}n - 2^3$$

    $$2\left(\frac{1}{3}n - 2^3\right) - 4 = 2\left(\frac{1}{3}n - 8\right) - 4 = \frac{2}{3}n - 16 - 4 = \frac{2}{3}n - 20$$

Practice Exercises 5.3B:

1. Translate "a number subtracted from the product of four and the number" into a variable expression. Then simplify.

2. Translate "three less than the sum of a number and twelve" into a variable expression. Then simplify.

3. Translate "a number plus the sume of the number and seven" into a variable expression. Then simplify.

4. Translate "the sum of two-fifths of a number and three-tenths of the number" into a variable expression. Then simplify.

5. Translate "three less than the total of a number and nine" into a variable expression. Then simplify.

6. Translate "four times the sum of five times a number and six" into a variable expression. Then simplify.

7. Translate "twelve more than the sum of ten and a number" into a variable expression. Then simplify.

8. Translate "three times the sum of two consecutive integers" into a variable expression. Then simplify.

9. Translate "one-half of the sum of two consecutive even integers" into a variable expression. Then simplify.

10. Translate "a number subtracted from the product of the number and five" into a variable expression. Then simplify.

Module 5: Variable Expressions
Objective 5.3C – Translate application problems

Many applications in mathematics require that you identify the unknown quantity, assign a variable to that quantity, and then attempt to express other unknown quantities in terms of the variable.

Example:

The length of a rectangle is 1 cm shorter than three times its width. Express the length of the rectangle in terms of the width.

Solution:

the width: w

1 cm shorter than three times its width: $3w-1$

Practice Exercises 5.3C

1. Ten gallons of paint were poured into two contrainers of different sizes. Use one variable to express the amount poured into each container.

2. The speed of a commercial jet is three times the speed of a corporate jet. Express the speed of the commercial jet in terms of the speed of the corporate jet.

3. The base of a triangle is 4 feet longer than the height of the triangle. Express the base of the triangle in terms of the height of the triangle.

4. The length of a rectangular piece of poster board is 4 times the width. Express the length of the paper in terms of the width.

5. A board 8 feet long was cut into two pieces. Express the length of the longer piece in terms of the length of the shorter piece.

6. The length of a rectangular area rug is 3 feet less than twice the width. Express the length of the rug in terms of the width of the rug.

7. The number of cherry trees in an orchard is approximately one-fifth the number of apple trees in the orchard. Express the number of cherry trees in the orchard in terms of the number of apple trees in the orchard.

8. The number of adjunct faculty at a community college is 206 more than the number of full-time faculty. Express the number of adjunct faculty in terms of the number of full-time faculty at the community college.

Answers

Practice Exercises 5.3A

1. $6 + x$
2. $12 - x$
3. $a - 1$
4. $\dfrac{z}{8}$
5. $x^2 + 10$
6. $5(n + 8)$
7. $x + 3x$
8. $-2t$
9. $5(y + 6)$
10. $\dfrac{1}{4}y^2 + 12$
11. $x + \dfrac{x}{2}$
12. $y - 2y$

Practice Exercises 5.3B

1. $4n - n$; $3n$
2. $(n + 12) - 3$; $n + 9$
3. $n + (n + 7)$; $2n + 7$
4. $\dfrac{2}{5}n + \dfrac{3}{10}n$; $\dfrac{7}{10}n$
5. $(n + 9) - 3$; $n + 6$
6. $4(5n + 6)$; $20n + 24$
7. $(n + 10) + 12$; $n + 22$
8. $3(n + n + 1)$; $6n + 3$
9. $\dfrac{1}{2}(n + n + 2)$; $n + 1$
10. $5n - n$; $4n$

Practice Exercises 5.3C

1. Gallons of paint in first container: g; gallons of paint in the second container: $10 - g$

2. Speed of corporate jet: s; speed of commercial jet: $3s$

3. Height of triangle: h; base of triangle: $h + 4$

4. Width: w; length: $4w$

5. Length of shorter piece: L; length of longer piece: $8 - L$

6. Width of rectangle: w; length of rectangle: $2w - 3$

7. Number of apples trees: N; number of cherry trees: $\dfrac{1}{5}N$

8. Number of full-time faculty: F; number of adjunct faculty: $F + 206$

6 Introduction to Equations

Module 6: Introduction to Equations

Objective 6.1A – Determine whether a given number is a solution of an equation

An **equation** expresses the equality of two mathematical expressions. A **solution of an equation** is a number that, when substituted for the variable, results in a true equation.

Examples:

1. Is -1 a solution of $4x - 2 = x^2 - 7$?

    Solution:

    $$4x - 2 = x^2 - 7$$

$4(-1) - 2$	$(-1)^2 - 7$
$-4 - 2$	$1 - 7$

    $$-6 = -6$$

    Yes, -6 is a solution of the equation.

2. Is 3 a solution of $5 - 2x = x - 2$?

    Solution:

    $$5 - 2x = x - 2$$

$5 - 2(3)$	$3 - 2$

    $$-1 \neq 1$$

    No, 3 is not a solution of the equation.

1. Is $\dfrac{1}{2}$ a solution of $x+\dfrac{3}{2}=2$?

2. Is -9 a solution of $3-x^2=-10x-6$?

3. Is -3 a solution of $3-z=z^2-3$?

Module 6: Introduction to Equations

Objective 6.1B – Solve an equation of the form $x + a = b$

To **solve** an equation means to find a solution of the equation. Equations that have the same solution are called **equivalent equations**. Adding the same number to each side of an equation produces an equivalent equation.

Addition Property of Equations

The same number can be added to each side of an equation without changing its solution. In symbols, the equation $a = b$ has the same solution as the equation $a + c = b + c$.

In solving an equation, the goal is to rewrite the given equation in the form *variable = constant*. The Addition Property of Equations is used to remove a term from one side of the equation by adding the opposite of that term to each side of the equation.

Because subtraction is defined in terms of addition, the Addition Property of Equations also makes it possible to subtract the same number from each side of an equation without changing the solution of the equation.

Examples:

1. Solve: $x + 4 = 2$

    Solution:

    $$x + 4 = 2$$
    $$x + 4 - 4 = 2 - 4$$
    $$x + 0 = -2$$
    $$x = -2$$

    - Subtract 4 from each side.
    - Simplify
    - The equation is of the form *variable = constant*.

    The solution is −2.

2. Solve: $\dfrac{2}{3} = z - \dfrac{1}{4}$

Solution:

$$\dfrac{2}{3} = z - \dfrac{1}{4}$$

$$\dfrac{2}{3} + \dfrac{1}{4} = z - \dfrac{1}{4} + \dfrac{1}{4}$$

$$\dfrac{8}{12} + \dfrac{3}{12} = z + 0$$

$$\dfrac{11}{12} = z$$

- Add $\dfrac{1}{4}$ to each side of the equation.
- Rewrite the fractions using a common denominator.
- The equation is of the form *constant = variable*.

The solution is $\dfrac{11}{12}$.

Practice Exercises 6.1B

1. Solve and check: $x + 3 = 8$

2. Solve and check: $a - 4 = 11$

3. Solve and check: $1 + a = 10$

4. Solve and check: $t + 8 = 0$

5. Solve and check: $x - 7 = -4$

6. Solve and check: $x + 4 = 4$

7. Solve and check: $z + 7 = 1$

8. Solve and check: $t - 5 = -3$

9. Solve and check: $9 + a = 15$

10. Solve and check: $-6 = n + 2$

11. Solve and check: $-9 = -3 + x$

12. Solve and check: $c + \dfrac{4}{5} = -\dfrac{1}{5}$

13. Solve and check: $x - \dfrac{1}{5} = \dfrac{2}{5}$

14. Solve and check: $x + \dfrac{1}{3} = -\dfrac{5}{6}$

Module 6: Introduction to Equations
Objective 6.1C – Solve an equation of the form $ax = b$

Multiplication Property of Equations

Each side of an equation can be multiplied by the same nonzero number without changing the solution of the equation. In symbols, if $c \neq 0,$, then the equation $a = b$ has the same solutions as the equation $ac = bc$.

The Multiplication Property of Equations is used to remove a coefficient by multiplying each side of the equation by the reciprocal of the coefficient.

Because division is defined in terms of multiplication, each side of an equation can be divided by the same nonzero number without changing the solution of the equation.

Examples:

1. Solve: $6 = 4x$

   Solution:

   $$6 = 4x$$
   $$\frac{6}{4} = \frac{4x}{4}$$

   - Divide each side by 4, the coefficient of x.

   $$\frac{3}{2} = x$$

   - The equation is of the form *constant = variable*.

   The solution is $\frac{3}{2}$.

2. Solve: $-\dfrac{3}{4}x = 9$

Solution:

$$-\dfrac{3}{4}x = 9$$

$$\left(-\dfrac{4}{3}\right)\left(-\dfrac{3}{4}x\right) = \left(-\dfrac{4}{3}\right)9$$ • Multiply each side by $-\dfrac{4}{3}$, the reciprocal of $-\dfrac{3}{4}$.

$$x = -12$$ • The equation is of the form *variable = constant*.

The solution is -12.

Practice Exercises 6.1C

1. Solve and check: $4x = 12$

2. Solve and check: $2a = -10$

3. Solve and check: $-6m = 18$

4. Solve and check: $-5n = -35$

5. Solve and check: $-42 = 6a$

6. Solve and check: $-\dfrac{y}{3} = 2$

7. Solve and check: $\dfrac{2}{3}y = 8$

8. Solve and check: $-\dfrac{3}{4}d = 9$

9. Solve and check: $\dfrac{2x}{7} = 4$

10. Solve and check: $-\dfrac{7z}{8} = 14$

11. Solve and check: $-9 = -\dfrac{3x}{4}$

12. Solve and check: $9x - 5x = 20$

Answers

Practice Exercises 6.1A

   1. Yes
   2. No
   3. Yes

Practice Exercises 6.1B

   1. 5
   2. 15
   3. 9
   4. -8
   5. 3
   6. 0
   7. -6
   8. 2
   9. 6
   10. -8
   11. -6
   12. -1
   13. $\dfrac{3}{5}$
   14. $-\dfrac{7}{6}$

Practice Exercises 6.1C

   1. 3
   2. -5
   3. -3
   4. 7
   5. -7
   6. -6
   7. 12
   8. -12
   9. 14
   10. -16
   11. 12
   12. 5

Module 6: Introduction to Equations
Objective 6.2A – Solve proportions

A **proportion** states the equality of two ratios or rates.

In the proportion $\dfrac{a}{b} = \dfrac{c}{d}$, the terms a and d are called the **extremes**; the terms b and c are called the **means**. In any true proportion, the product of the means equals the product of the extremes. This is sometimes phrased as "the cross products are equal."

Example:

Determine whether the proportion $\dfrac{18 \text{ mi}}{10 \text{ gal}} = \dfrac{40.5 \text{ mi}}{22.5 \text{ gal}}$ is a true proportion.

Solution:

The product of the means: $(10)(40.5) = 405$
The product of the extremes: $(18)(22.5) = 405$

The proportion is true because $405 = 405$.

When three terms of a proportion are given, the fourth term can be found. To solve a proportion for an unknown term, use the fact that the product of the means equals the product of the extremes.

Practice Exercises 6.2A

1. Determine if the proportion is true or not true: $\dfrac{6}{7} = \dfrac{12}{14}$

2. Determine if the proportion is true or not true: $\dfrac{9}{18} = \dfrac{12}{24}$

3. Determine if the proportion is true or not true:

$$\dfrac{606 \text{ words}}{10 \text{ minutes}} = \dfrac{302 \text{ words}}{5 \text{ minutes}}$$

4. Determine if the proportion is true or not true:

$$\dfrac{48 \text{ cents}}{4 \text{ hours}} = \dfrac{60 \text{ cents}}{5 \text{ hours}}$$

5. Determine if the proportion is true or not true:

$$\frac{450 \text{ gallons}}{60 \text{ minutes}} = \frac{180 \text{ gallons}}{24 \text{ minutes}}$$

6. Determine if the proportion is true or not true:

$$\frac{6{,}200 \text{ words}}{40 \text{ pages}} = \frac{7{,}750 \text{ words}}{50 \text{ pages}}$$

7. Solve. Round to the nearest hundredth: $\dfrac{n}{8} = \dfrac{20}{32}$

8. Solve. Round to the nearest hundredth: $\dfrac{8}{n} = \dfrac{9}{27}$

9. Solve. Round to the nearest hundredth: $\dfrac{n}{8} = \dfrac{9}{12}$

10. Solve. Round to the nearest hundredth: $\dfrac{35}{n} = \dfrac{22}{11}$

11. Solve. Round to the nearest hundredth: $\dfrac{n}{15} = \dfrac{0.8}{5.6}$

12. Solve. Round to the nearest hundredth: $\dfrac{1.8}{18} = \dfrac{n}{12}$

Module 6: Introduction to Equations
Objective 6.2B – Solve application problems using proportions

In setting up a proportion, keep the same units in the numerators and the same units in the denominators.

Example:

Suppose that 2 gallons of paint can cover 580 square feet of wall space. How many square feet of wall space can 7 gallons cover?

Solution:

Let n represent the number of square feet of wall space that 7 gallons can cover.

$$\frac{580 \text{ sq ft}}{2 \text{ gal}} = \frac{n \text{ sq ft}}{7 \text{ gal}}$$

$$7 \cdot 580 = 2n$$

$$\frac{4{,}060}{2} = \frac{2n}{2}$$

$$2{,}030 = n$$

7 gallons of paint can cover 2,030 sq ft of wall space.

Practice Exercises 6.2B
Solve. Round to the nearest hundredth.

1. A life insurance policy costs $4.25 for every $1,000 of insurance. At this rate, what is the cost for $20,000 worth of life insurance?

2. A liquid plant food is prepared by using one gallon of water for each 1.5 teaspoon of plant food. At this rate, how many teaspoons of plant food are required for 7 gallons of water?

3. A $19.75 sales tax is charged for a $395 purchase. At this rate, what is the sales tax for a $621 purchase?

4. The scale on the plan for a new office building is 1 inch equals 4 feet. How long is room that measures $8\frac{1}{2}$ inches on the drawing?

Solve. Round to the nearest hundredth.

5. A stock investment of 150 shares paid a dividend of $555. At this rate, what dividend would be paid on 280 shares of stock?

6. A bank demands a loan payment of $18,95 each month for every $1,000 borrowed. At this rate, what is the monthly payment for a $6,000 loan?

7. For every 10 people who work in a city, 7 of them commute by public transportation. If 34,600 people work in the city, how many of them do not take public transportation?

8. For every 15 gallons of water pumped into the holding tank, 8 gallons were pumped out. After 930 gallons had been pumped in, how much water remained in the tank?

Answers

Practice Exercises 6.2A

1. True
2. True
3. Not true
4. True
5. True
6. True
7. 5
8. 24
9. 6
10. 17.5
11. 2.14
12. 1.2

Practice Exercises 6.2B

1. $85.00
2. 10.5 teaspoons
3. $31.05
4. 34 feet
5. $1,036
6. $113.70
7. 10,380 people
8. 434 gallons

Module 6: Introduction to Equations
Objective 6.3A – Solve the basic percent equation

The Basic Percent Equation

Percent · base = amount

Look for the number or phrase that follows the word *of* when determining the base in the basic percent equation.

In most cases, the percent is written as a decimal before the basic percent equation is solved. However, some percents are more easily written as a fraction than as a decimal.

Examples:

1. What is 16% of 30?

   Solution:

   Percent · base = amount
   $$0.16 \cdot 30 = n$$
   $$4.8 = n$$

   16% of 30 is 4.8.

2. 45% of what number is 18?

   Solution:

   Percent · base = amount
   $$0.45 \cdot n = 18$$
   $$\frac{0.45n}{0.45} = \frac{18}{0.45}$$
   $$n = 40$$

   45% of 40 is 18.

3. What percent of 5 is 4?

Solution:

$$\text{Percent} \cdot \text{base} = \text{amount}$$
$$n \cdot 5 = 4$$
$$\frac{5n}{5} = \frac{4}{5}$$
$$n = 0.80 = 80\%$$

80% of 5 is 4.

Practice Exercises 6.3A
Use the percent equation to solve each.

1. 8% of 45 is what?

2. 15% of 100 is what?

3. What is 45% of 60?

4. What is 65% of 135.5?

5. What is 0.02% of 250?

6. 5% of 900 is what?

7. 150% of 98 is what?

8. Find 12% of 540.

9. Find 0.08% of 375.

10. What is $6\frac{1}{4}$% of 620?

Module 6: Introduction to Equations
Objective 6.3B – Solve percent problems using proportions

The proportion method is based on writing two ratios. One ratio is the percent ratio, written as $\dfrac{\text{percent}}{100}$. The second ratio is the amount-to-base ratio, written as $\dfrac{\text{amount}}{\text{base}}$. These two ratios form the proportion

$$\frac{\text{percent}}{100} = \frac{\text{amount}}{\text{base}}$$

Examples

1. 126 is 84% of what number?

   Solution:

   $$\frac{\text{percent}}{100} = \frac{\text{amount}}{\text{base}}$$
   $$\frac{84}{100} = \frac{126}{n}$$
   $$84 \cdot n = 100 \cdot 126$$
   $$\frac{84n}{84} = \frac{12{,}600}{84}$$
   $$n = 150$$

   126 is 84% of 150.

2. Find 14% of 70.

   Solution:

   $$\frac{\text{percent}}{100} = \frac{\text{amount}}{\text{base}}$$
   $$\frac{14}{100} = \frac{n}{70}$$
   $$100 \cdot n = 14 \cdot 70$$
   $$\frac{100n}{100} = \frac{980}{100}$$
   $$n = 9.8$$

   14% of 70 is 9.8

Practice Exercises 6.3B
Use the proportion method to solve each.

1. 34% of 850 is what?

2. 54 is what percent of 180?

3. 90% of what is 54?

4. What percent of 324 is 162?

5. What is 160% of 480?

6. 27 is 45% of what?

7. What percent of 110 is 44?

8. What percent of 220 is 11?

Module 6: Introduction to Equations
Objective 6.3C – Solve application problems

Examples:

1. An 18-carat yellow-gold necklace contains 75% gold, 16% silver, and 9% copper. If the necklace weighs 25 grams, how many grams of copper are in the necklace?

    Solution:

    To find the number of grams of copper in the necklace, write and solve the basic percent equation using n to represent the amount of copper. The percent is 9% and the base is 25 grams.

    $$9\% \cdot 25 = n$$
    $$0.09 \cdot 25 = n$$
    $$2.25 = n$$

    2.25 grams of copper are in the necklace.

2. In a survey, 1,236 adults nationwide were asked, "What irks you most about the actions of other motorists?" 293 people gave the response "tailgaters." What percent of those surveyed were most irked by tailgaters? Round to the nearest tenth of a percent.

    Solution:

    To find what percent were most irked by tailgaters, write and solve the basic percent equation using n to represent the unknown percent. The base is 1,236 and the amount is 293.

    $$1{,}236n = 293$$

    $$\frac{1{,}236n}{1{,}236} = \frac{293}{1{,}236}$$

    $$n \approx 0.237$$

    Approximately 23.7% of those surveyed were most irked by tailgaters.

Practice Exercises 6.3C

1. A nursery sold 450 geranium plants in June. In July, the nursery increased its sales by 8%. How many geranium plants were sold in July?

2. An office building has an appraised value of $8,000,000. The real estate taxes are 2.35% of the appraised value of the building. Find the real estate taxes.

3. A used car salesperson sold 6 of the 50 cars in the lot. What percent of the total number of cards in the lot were sold?

4. A survey of 1,760 people showed that 352 people favored the incumbent mayor. What percent of the people surveyed favored the the incumbent mayor?

5. A total of $4,200 was paid in taxes on an income of $16,800. Find the percent of the total income paid in taxes.

6. A soccer team won 51 out of 68 games they played. What percent of the games played did they win?

7. A During shipping, 600 of the 4,800 light bulbs were damaged. What percent of the number of light bulbs were damaged in the shipping?

8. An administrative assistant types 70 words per minutes with 98% accuracy. During 5 minutes of typing, how many errors does the secretary make?

Answers

Practice Exercises 6.3A

1. 3.6
2. 15
3. 27
4. 88.075
5. 0.05
6. 45
7. 147
8. 64.8
9. 0.3
10. 38.75

Practice Exercises 6.3B

1. 289
2. 30%
3. 60
4. 50%
5. 768
6. 60
7. 40%
8. 5%

Practice Exercises 6.3C

1. 486 plants
2. $188,000
3. 12%
4. 20%
5. 25%
6. 75%
7. 12.5%
8. 7 errors

Module 6: Introduction to Equations
Objective 6.4A – Solve percent increase problems

Percent increase is used to show how much a quantity has increased over its original value.

To find the percent increase:

- Find the amount of increase.
- Use the basic percent equation, where the amount is the amount of increase.

Example:

A car's gas mileage increases from 17.5 miles per gallon to 18.2 miles per gallon. Find the percent increase in gas mileage.

Solution:

The amount of increase is $18.2 - 17.5 = 0.7$.

$$\text{Percent} \cdot \text{base} = \text{amount}$$
$$n \cdot 17.5 = 0.7$$
$$\frac{17.5n}{17.5} = \frac{0.7}{17.5}$$
$$n = 0.04 = 4\%$$

The percent increase in gas mileage is 4%.

Practice Exercises 6.4A

Solve. Round to the nearest tenth of a percent or to the nearest cent.

1. The value of a $7,000 investment increased $1,750. What percent increase does this represent?

2. A sweater which sold for $24 last month increase in price by $2. What percent increase does this represent?

3. The number of students enrolled in a speed reading course increased from 60 to 66 during the first 10 days of school. What is the percent increase?

4. A manufacturer of ceiling fans increased its monthly output of 1,500 by 10%. What is the amount of increase?

Solve. Round to the nearest tenth of a percent or to the nearest cent.

5. The employees of a manufacturing plant received a 6% increase in pay.

    a. What is the amount of the increase for an employee who makes $225 per week?

    b. What is the weekly wage for the employee after the wage increase?

6. A college increase its number of parking spaces from 1,000 to 1,050.
    a. How many new spaces were added?
    b. What percent increase does this represent?

7. A cafeteria increased the number of items on the menu from 80 to 90. What percent of increase does this represent?

Module 6: Introduction to Equations
Objective 6.4B – Solve percent decrease problems

Percent decrease is used to show how much a quantity has decreased from its original value.

To find the percent decrease:

- Find the amount of decrease.
- Use the basic percent equation, where the amount is the amount of decrease.

Example:

A new production method reduced the time needed to clean a piece of metal from 8 min to 5 min. What percent decrease does this represent?

Solution:

The amount of decrease is $8 - 5 = 3$.

$$\text{Percent} \cdot \text{base} = \text{amount}$$
$$n \cdot 8 = 3$$
$$\frac{8n}{8} = \frac{3}{8}$$
$$n = 0.375 = 37.5\%$$

The markup rate (percent increase) is 37.5%.

Practice Exercises 6.4B

1. A health spa sold 120 memberships in November. In May the spa sold 18 fewer memberships than in November. What was the percent decrease in the number of memberships sold?

2. By washing all the clothes in cold water, a family was able to reduce its normal mothly utility bill of $125 by $15. What percent decrease does this represent?

3. It is estimated that the value of a motocycle is reduced by 25% after one year of ownership. Using this estimate, how much value does a $1,500 new motocycle lose after one year?

4. Because of a decrease in orders for telephones, a telephone center reduced the orders for phones in from 140 per month to 91 per month.
   a. What is the amount of decrease?
   b. What percent decrease does this represent?

5. Last year a company earned a profit of $285,000. This year, the company's profits were 6% less than last year's,
    a. What was the amount of decrease?
    b. What was the profit of this year?

6. The price of a new model digital camera dropped from $150 to $114 in ten months. What percent decrease does this represent?

Answers

Practice Exercises 6.4A

1. 25%
2. 8.3%
3. 10%
4. 150 fans
5a. $13.50
5b. $238.50
6a. 50 spaces
6b. 5%
7. 12.5%

Practice Exercises 6.4B

1. 15%
2. 12%
3. $375
4a. 49 orders
4b. 35%
5a. $17,100
5b. $267,900
6. 24%

Module 6: Introduction to Equations
Objective 6.5A – Solve markup problems

Cost is the amount a merchandising business or retailer pays for a product. Selling price, or retail price, is the price for which a merchandising business or retailer sells a product to a customer.

The difference between selling price and cost is called **markup**.

Markup Equations

$M = P - C$

$P = C + M$

$M = r \cdot C$

where M = markup, P = selling price, C = cost, and r = markup rate.

Examples:

1. A watch that cost $98 is being sold for $156.80. Find the markup rate.

   Solution:

   $M = S - C$

   $M = 156.80 - 98$

   $M = 58.80$

   $M = r \cdot C$

   $58.80 = r \cdot 98$

   $\dfrac{58.80}{98} = \dfrac{98r}{98}$

   $0.60 = r$

   The markup rate is 60%.

2. A video game costing $47 has a markup rate of 75%. Find the selling price.

   Solution:

   $M = r \cdot C$

   $M = 0.75 \cdot 47$

   $M = 35.25$

$$P = C + M$$
$$P = 47 + 35.25$$
$$P = 82.25$$

The selling price is $82.25.

Practice Exercises 6.5A

1. A store manager used a markup rate of 30% on all desk lamps. What is the markup on a lamp which costs the store $26?

2. An automobile tire dealer uses a markup rate of 32%. What is the markup on tires which cost the dealer $34?

3. A beach-wear shop uses a markup rate of 40% on a bathing suit which costs the shop $34.
   a. What is the markup?
   b. What is the selling price?

4. A produce market pays $1.09 for pineapples. The market uses a 55% markup rate.

    a. What is the markup?

    b. What is the selling price?

5. A store uses a markup rate of 38%. What is the selling price for the DVD player which costs the store $47?

Module 6: Introduction to Equations
Objective 6.5B – Solve discount problems

The **discount**, or **markdown**, is the amount by which a retailer reduces the regular price of a product. The percent discount is called the **discount rate** and is usually expressed as a percent of the original price (the regular selling price).

Discount Equations

$D = R - S$

$D = r \cdot R$

$S = (1-r)R$

where D = discount or markdown, S = sale price, R = regular price, and r = discount rate.

Examples:

1. Find the markdown rate on a round-trip airfare from New York to Paris that has a regular price of $1,295 and is on sale for $995. Round to the nearest tenth of a percent.

   Solution:

   $D = R - S$
   $D = 1,295 - 995$
   $D = 300$

   $D = r \cdot S$
   $300 = r \cdot 1,295$
   $\dfrac{300}{1,295} = \dfrac{1,295r}{1,295}$
   $0.232 \approx r$

   The markdown rate is approximately 23.2%.

2. A gold ring with a regular price of $415 is on sale for 55% off the regular price. Find the sale price.

   Solution:

   $$S = (1-r)R$$
   $$S = (1-0.55)415$$
   $$S = 0.45 \cdot 415$$
   $$S = 186.75$$

   The sale price is $186.75.

Practice Exercises 6.5B

1. To promote business, a store manager offers a small vacuum cleaner which regularly sells for $25 at $9 off the regular price. What is the discount rate?

2. A sporting goods store is selling its $150 exercise bike for 20% off the regular price. What is the discount?

3. A hardware store is selling its $32 lock set for 15% off the regular price. What is the discount?

4. A gift shop has its picture frames which regularly cost $35 on sale $30.80.
   a. What is the discount?
   b. What is the discount rate?

5. An automobile body shop has regularly priced $600 paint jobs on sale for $480.
   a. What is the discount?
   b. What is the discount rate?

6. During a going-out-of-business sale, all lawn and garden merchandise was reduced 40% off the regular price. What was the sale price of a lawn mower which normally sells for $230?

Answers

Practice Exercises 6.5A

1. $7.80
2. $10.88
3a. $13.60
3b. $47.60
4a. $0.60
4b. $1.69
5. $64.86

Practice Exercises 6.5B

1. 36%
2. $30
3. $4.80
4a. $4.20
4b. 12%
5a. $120
5b. 20%
6. $138

Module 6: Introduction to Equations
Objective 6.6A – Solve simple interest problems

The amount you deposit in the savings account is called the **principal**. The amount the bank pays you for the privilege of using the money is called **interest**.

If you borrow money from the bank in order to buy a car, the amount you borrow is called the **principal**. The additional amount of money you must pay the bank, above and beyond the amount borrowed, is called **interest**.

The percent used to determine the amount of interest to be paid is the **interest rate**.

Interest computed on the original principal is called simple interest.

The Simple Interest Formula
$I = Prt$, where I = simple interest earned, P = principal,
r = annual simple interest rate, and t = time (in years)

Examples:

1. A rancher borrowed $120,000 for 180 days at an annual simple interest rate of 8.75%. What is the simple interest due on the loan?

   Solution:

   Because an annual interest rate is given, the time must be converted from days to years. $180 \text{ days} = \dfrac{180}{365} \text{ year}$

   $I = Prt$

   $I = 120,000 \cdot 0.0875 \cdot \dfrac{180}{365}$

   $I \approx 5,178.08$

   The simple interest due on the loan is $5,178.08.

2. A $12,000 investment earned $462 in 6 months. Find the annual simple interest rate on the loan.

Solution:

Because an annual interest rate is given, the time must be converted from months to years. $6 \text{ months} = \dfrac{6}{12} \text{ year}$

$$I = Prt$$
$$462 = 12{,}000r\left(\dfrac{1}{2}\right)$$
$$462 = 6{,}000r$$
$$\dfrac{462}{6{,}000} = \dfrac{6{,}000r}{6{,}000}$$
$$0.077 = r$$

The annual simple interest rate is 7.7%.

The principal plus the interest owed on a loan is called the **maturity value**.

Formula for the Maturity Value of a Simple Interest Loan
$M = P + I$, where $M =$ the maturity value, $P =$ the principal, and
$I =$ the simple interest

Example:

You arrange for a 4-month bank loan of $8,000 at an annual simple interest rate of 9%. Find the maturity value of the loan.

Solution:

$$I = Prt$$
$$I = 8{,}000(0.09)\dfrac{4}{12}$$
$$I = 240$$

$$M = P + I$$
$$M = 8{,}000 + 240$$
$$M = 8{,}240$$

The maturity value of the loan is $8,240.

Practice Exercises 6.6A

1. To finance the purchase of 8 new taxicabs, the owners of the fleet borrows $84,000 for 8 months at an annual interest rate of 16%. What is the simple interest due on the loan?

2. An executive was offered a $34,000 loan at a 14.5% annual interest rate for three years. Find the simple interest due on the loan.

3. You arrange for a 6-month bank loan of $6,000 at an annual simple rate of 7.5%. Find the total amount you must repay to the bank.

4. A bank charges its customers an interest rate of 1.8% per month for transferring money into an account which is overdrawn. Find the interest owed to the bank for one month when $300 was transferred into an overdrawn account.

5. A copier is purchased and a $2,100 loan is obtained for two years at a simple interest rate of 17%.
    a. Find the interest due on the loan.
    b. Find the maturity value of the loan.

6. To reduce their inventory of new cars, a dealer is offering car loans at a simple annual interest rate of 11%.

    a. Find the interest charged to a customer who financed a car loan of $8,200 for four years.
    b. Find the maturity value of the loan.

Answers

Practice Exercises 6.6A

1. $8,960
2. $14,790
3. $6,225
4. $5.40
5a. $714
5b. $2,814
6a. $3,608
6b. $11,808

MODULE

8

Linear Functions and Inequalities in Two Variables

Module 8: Linear Functions and Inequalities in Two Variables

Objective 8.1A – Find the length and midpoint of a line segment

A **rectangular coordinate system** is formed by two number lines, one horizontal and one vertical, that intersect at the zero point of each line. The point of intersection is called the **origin**. The two lines are called **coordinate axes**, or simply **axes**.

The axes determine a **plane**. The two axes divide the plane into four regions called **quadrants**, numbered counterclockwise from I to IV.

Each point in the plane can be identified by a pair of numbers called an **ordered pair**. The **coordinates** of a point are the numbers in the ordered pair associated with the point. The **graph of an ordered pair** is the dot drawn at the coordinates of the point in the plane.

When drawing a rectangular coordinate system, we often label the horizontal axis x and the vertical axis y. In this case, the coordinate system is called an *xy*-**coordinate system**. The coordinates of the points are given by ordered pairs (x, y), where the first number is the *x*-**coordinate** and the second number is the *y*-**coordinate**.

For example, in the diagram below, point A has coordinates $(-3, 4)$, and point B has coordinates $(4, -3)$.

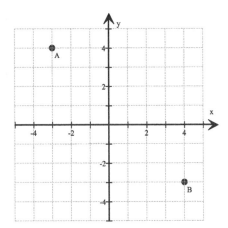

Practice Exercises 8.1A:

1. What quadrant is $(-3, 5)$ in?

2. What quadrant is $(1, -4)$ in?

3. Plot the following points:

A. $(2, 4)$

B. $(0, -3)$

C. $(-1, 2)$

D. $(-5, 0)$

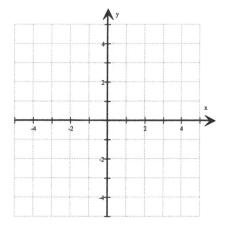

4. Plot the following points:

A. $(-2, -4)$

B. $(2, -3)$

C. $(1, 2)$

D. $(-1, 0)$

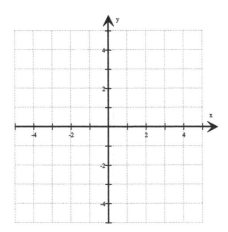

Answers

Practice Exercises 8.1A

1. Quadrant II

2. Quadrant IV

3.

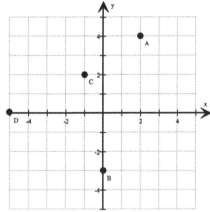

4.

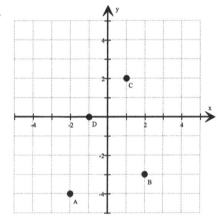

Module 10: Polynomials
Objective 10.2A – Multiply monomials

Rule for Multiplying Exponential Expressions
If m and n are integers, then $x^m \cdot x^n = x^{m+n}$.

Examples:

1. Multiply: $\left(3x^2y\right)\left(-4xy\right)$

   Solution:

   $$\left(3x^2y\right)\left(-4xy\right) = 3\left(-4\right)x^{2+1}y^{1+1} = -12x^3y^2$$

2. Multiply: $\left(-xy\right)\left(-yz^2\right)\left(4x\right)$

   Solution:

   $$\left(-xy\right)\left(-yz^2\right)\left(4x\right) = \left(-1\right)\left(-1\right)\left(4\right)x^{1+1}y^{1+1}z^2 = 4x^2y^2z^2$$

Practice Exercises 10.2A

1. Multiply: $\left(x^5\right)\left(x^7\right)$

2. Multiply: $\left(x^0y^2\right)\left(y^3\right)$

3. Multiply: $(x^3 y)(x^2 y^3)$

4. Multiply: $(3ab)(-5a^6 b)$

5. Multiply: $\dfrac{3}{4}x^{10} \cdot 20xz^4$

6. Multiply: $(-rst^3)(3r^2 st)$

Module 10: Polynomials
Objective 10.2B – Simplify powers of monomials

Rule for Simplifying Powers of Exponential Expressions

If m and n are integers, then $\left(x^m\right)^n = x^{mn}$.

Rule for Simplifying Powers of Products

If m, n, and p are integers, then $\left(x^m y^n\right)^p = x^{mp} y^{np}$.

Example:

Simplify: $\left(-2x^2 y\right)\left(-4x^2 yz\right)^3$

Solution:

$$\left(-2x^2 y\right)\left(-4x^2 yz\right)^3 = \left(-2x^2 y\right)\left(-4\right)^3 \left(x^2\right)^3 y^3 z^3$$
$$= \left(-2x^2 y\right)\left(-64\right) x^6 y^3 z^3$$
$$= \left(-2\right)\left(-64\right) x^{2+6} y^{1+3} z^3$$
$$= 128 x^8 y^4 z^3$$

Practice Exercises 10.2B

1. Simplify: $\left(x^4\right)^5$

2. Simplify: $\left(a^3 b\right)^2$

3. Simplify: $\left(-3yz^5\right)^0$

4. Simplify: $\left(2x^4y^2\right)^3$

5. Simplify: $\left(-yz^5\right)^6$

6. Simplify: $\left(2ab\right)^2\left(-3a^2b\right)$

Answers

Practice Exercises 10.2A

1. x^{12}
2. y^5
3. $x^5 y^4$
4. $-15a^7 b^2$
5. $15x^{11} z^4$
6. $-3r^3 s^2 t^4$

Practice Exercises 10.2B

1. x^{20}
2. $a^6 b^2$
3. 1
4. $8x^{12} y^6$
5. $y^6 z^{30}$
6. $-12a^4 b^3$

Module 19: Measurement
Objective 19.1A – Convert units of length in the U.S. Customary System

A **measurement** includes a number and a unit. The standard U.S. Customary units of **length**, or distance, are **inch**, **foot**, **yard**, and **mile**. The list below gives equivalences between some units of length.

Equivalences Between Units of Length in the U.S. Customary System
12 inches (in.) = 1 foot (ft)
3 ft = 1 yard (yd)
5,280 ft = 1 mile (mi)

A **conversion rate** is used to change from one unit to another. **Dimensional analysis** involves using conversion rates to change from one unit of measurement to another unit of measurement.

Choose a conversion rate so that the unit in the numerator of the conversion rate is the same as the unit needed in the answer. The unit in the denominator of the conversion rate is the same as the unit in the given measurement.

Example:

Convert 93 ft to yards.

Solution:

The conversation rate $\frac{1 \text{ yd}}{3 \text{ ft}}$ is chosen so that the numerator unit, yards, is the same as the unit needed in the answer. The denominator unit, feet, is the same as the unit given in the measurement. Notice that since 1 yard = 3 feet, the conversion rate equals 1.

$$93 \text{ ft} = 93 \text{ ft} \times \frac{1 \text{ yd}}{3 \text{ ft}} = \frac{93}{3} \text{ yd} = 31 \text{ yd}$$

$$93 \text{ ft} = 31 \text{ yd}$$

If you do not know the direct conversion rate to change from one unit to another, a conversion can be accomplished by multiplying by two or more conversion rates.

Example:

Convert $2\frac{3}{4}$ yd to inches.

Solution:

Based on the equivalence table above we can convert yards to feet and then feet to inches.

$$2\frac{3}{4}\text{ yd} = 2\frac{3}{4}\ \cancel{\text{yd}} \times \frac{3\ \cancel{\text{ft}}}{1\ \cancel{\text{yd}}} \times \frac{12\text{ in.}}{1\ \cancel{\text{ft}}} = 2\frac{3}{4} \times 3 \times 12\text{ in.}$$

$$= \frac{11}{4} \times 3 \times 12\text{ in.}$$

$$= 99\text{ in.}$$

$$2\frac{3}{4}\text{ yd} = 99\text{ in.}$$

Practice Exercises 19.1A:

For Exercises 1 through 6, convert between the two measurement units.

1. 108 in. = _____ ft

2. $1\frac{1}{2}$ mi = _____ yd

3. 15,840 yd = _____ mi

4. 441 in. = _____ yd

5. $2\frac{1}{2}$ yd = _____ in

6. $1\frac{1}{4}$ mi = _____ ft

Module 19: Measurement

Objective 19.1B – Convert units of weight in the U.S. Customary System

Weight is a measure of how strongly Earth is pulling on an object. The U.S. Customary System units of weight are ounce, pound, and ton.

The equivalences shown below can be used to form conversion rates to change one unit of weight to another.

Equivalences Between Units of Weight in the U.S. Customary System
16 ounces (oz) = 1 pound (lb)   2,000 lb = 1 ton

Examples:

1. Convert 13,500 lb to tons.

    Solution:

    $$13,500 \text{ lb} = 13,500 \ \cancel{\text{lb}} \times \frac{1 \text{ ton}}{2,000 \ \cancel{\text{lb}}} = \frac{13,500}{2,000} \text{ ton} = 6\frac{3}{4} \text{ ton}$$

    $$13,500 \text{ lb} = 6\frac{3}{4} \text{ ton}$$

2. Convert $2\frac{1}{4}$ tons to ounces.

    Solution:

    $$2\frac{1}{4} \text{ tons} = 2\frac{1}{4} \ \cancel{\text{tons}} \times \frac{2,000 \ \cancel{\text{lb}}}{1 \ \cancel{\text{ton}}} \times \frac{16 \text{ oz}}{1 \ \cancel{\text{lb}}}$$

    $$= 2\frac{1}{4} \times 2,000 \times 16 \text{ oz}$$

    $$= 72,000 \text{ oz}$$

    $$2\frac{1}{4} \text{ tons} = 72,000 \text{ oz}$$

Practice Exercises 19.1B:

For Exercises 1 through 6, convert between the two measurement units.

1. 5 lb = _____ oz

2. 19,000 tons = _____ lb

3. 100 oz = _____ lb

4. 120,000 oz = _____ tons

5. 4,400 lb = _____ tons

6. $1\frac{1}{4}$ tons = _____ lb

Module 19: Measurement

Objective 19.1C – Convert units of capacity in the U.S. Customary System

Liquid substances are measured in units of **capacity**. The standard U.S. Customary units of capacity are the **fluid ounce**, **cup**, **pint**, **quart**, and **gallon**.

The equivalences shown below can be used to form conversion rates to change one unit of capacity to another.

Equivalences Between Units of Capacity in the U.S. Customary System
8 fluid ounces (fl oz) = 1 cup (c)
2 c = 1 pint (pt)
2 pt = 1 quart (qt)
4 qt = 1 gallon (gal)

Example:

    Convert 96 c to gallons.

Solution:

$$96 \text{ c} = 96 \text{ c} \times \frac{1 \text{ pt}}{2 \text{ c}} \times \frac{1 \text{ qt}}{2 \text{ pt}} \times \frac{1 \text{ gal}}{4 \text{ qt}} = \frac{96}{16} \text{ gal} = 6 \text{ gal}$$

$$96 \text{ c} = 6 \text{ gal}$$

Practice Exercises 19.1C:

For Exercises 1 through 6, convert between the two measurement units.

    1. 36 fl oz = _____ c

2. $2\frac{1}{2}$ gal = _____ qt

3. 8 pt = _____ fl oz

4. $8\frac{1}{2}$ qt = _____ c

5. 42 c = _____ qt

6. $7\frac{1}{2}$ pt = _____ qt

Module 19: Measurement
Objective 19.1D – Convert units of time

Some units in which time is measured are the **second**, **minute**, **hour**, **day**, and **week**.

The equivalences shown below can be used to form conversion rates to change one unit of time to another.

Equivalences Between Units of Time
60 seconds (s) = 1 minute (min)
60 min = 1 hour (h)
24 h = 1 day
7 days = 1 week

Example:

Convert 90,000 minutes to days.

Solution:

$$90,000 \text{ min} = 90,000 \text{ min} \times \frac{1 \text{ h}}{60 \text{ min}} \times \frac{1 \text{ day}}{24 \text{ h}}$$

$$= \frac{90,000}{1440} \text{ days}$$

$$= 62\frac{1}{2} \text{ days}$$

$$90,000 \text{ mi} = 62\frac{1}{2} \text{ days}$$

Practice Exercises 19.1D:

For Exercises 1 through 5, convert between the two measurement units.

1. $3\frac{2}{3}$ h = _____ min

2. 574 days = _____ weeks

3. 31,500 s = _____ h

4. $4\dfrac{1}{2}$ days = _____ min

5. 1,344 h = _____ weeks

Answers

Practice Exercises 19.1A

1. 9
2. 2,640
3. 9
4. $12\frac{1}{4}$
5. 90
6. 6,600

Practice Exercises 19.1B

1. 80
2. $9\frac{1}{2}$
3. $6\frac{1}{4}$
4. $3\frac{3}{4}$
5. $2\frac{1}{5}$
6. 2,500

Practice Exercises 19.1C

1. $4\frac{1}{2}$
2. 10
3. 128
4. 34
5. $10\frac{1}{2}$
6. $3\frac{3}{4}$

Practice Exercises 19.1D

1. 220
2. 82
3. $8\frac{3}{4}$
4. 6,480
5. 8

MTH 086 – Introductory Algebra
Course Content Outline
Course Outline Revision Date – January 2017

Based on the text, **Mathematics: Journey from Basic Mathematics through Intermediate Algebra**, 1st ed. by Aufmann and Lockwood, published by Cengage Learning

This schedule is subject to change. Instructor may announce changes at any time.

Class Meeting (80 mins)	Class Meeting (125 mins)	Module/Section	
		MODULE 1: WHOLE NUMBERS	
1	1	1.1	Introduction to Whole Numbers (Obj. A, B, C)
		1.2	Addition and Subtraction of Whole Numbers (Obj. A, B, C)
2		1.3	Multiplication and Division of Whole Numbers (Obj. A, B)
3	2	1.3	Multiplication and Division of Whole Numbers (Obj. C)
4		1.4	Exponential Notation and Order of Operations Agreement (Obj. A, B)
		MODULE 2: INTEGERS	
5	3	2.1	Introduction to Integers (Obj. A, B)
6		2.2	Addition and Subtraction of Integers (Obj. A, B)
7		2.2	Addition and Subtraction of Integers (Obj. C)
8	4	2.3	Multiplication and Division of Integers (Obj. A, B, C)
9		2.4	Exponents and the Order of Operations Agreement (Obj. A, B)
10	5	* Summary and Review for Test #1	
11	6	**Test #1** on Chapters 1 and 2	
		MODULE 3: FRACTIONS	
12	7	3.1	Least Common Multiple and Greatest Common Factor (Obj. A, B, C)
13		3.2	Introduction to Fractions (Obj. A, B)
14	8	3.3	Writing Equivalent Fractions (Obj. A, B, C)
15		3.4	Multiplication and Division of Fractions (Obj. A, B, C)
16	9	3.5	Addition and Subtraction of Fractions (Obj. A, B, C)
17	10	3.6	Operations on Positive and Negative Fractions (Obj. A, B)
18		3.7	The Order of Operations Agreement and Complex Fractions (Obj. A, B)
19	11	* Summary and Review for the Midterm Exam	
20	12	**Midterm Exam** on Chapters 1 through 3	

Class Meeting (80 mins)	Class Meeting (125 mins)	Module/Section
		MODULE 4: DECIMALS AND PERCENTS
21	13	4.1 Introduction to Decimals (Obj. A, B, C)
22		4.2 Addition and Subtraction of Decimals (Obj. A, B)
23	14	4.3 Multiplying and Dividing Decimals (Obj. A, B)
24		4.3 Multiplying and Dividing Decimals (Obj. B, C)
25	15	4.4 Comparing and Converting Fractions and Decimals (Obj. A, B, C)
		4.6 Radical Expressions and Real Numbers (Obj. A)
		MODULE 5: VARIABLE EXPRESSIONS
26	16	5.2 Simplifying Variable Expressions (Obj. A, B, C, D)
27		5.3 Translating Verbal Expressions into Variable Expressions (Obj. A, B, C)
		MODULE 10: POLYNOMIALS
28		10.2 Multiplication of Monomials (Obj. A, B)
		MODULE 6: INTRODUCTION TO EQUATIONS
29	17	6.1 Introduction to Equations (Obj. A, B, C)
		MODULE 8: LINEAR FUNCTIONS AND INEQUALITIES IN TWO VARIABLES
30		8.1 The Rectangular Coordinate System (Obj. A)
31	18	* Summary and Review for Test #2
32	19	**Test #2** on Chapters 4, 5, and 6
		MODULE 4: DECIMALS AND PERCENTS
33	20	4.4 Comparing and Converting Fractions and Decimals (Obj. D)
		MODULE 19: MEASUREMENT
34		19.1 The U.S. Customary System of Measurement (Obj. A, B, C, D)
		MODULE 6: INTRODUCTION TO EQUATIONS
35	21	6.2 Proportion (Obj. A, B)
		MODULE 4: DECIMALS AND PERCENTS
36	22	4.5 Introduction to Percents (Obj. A, B)

Class Meeting (80 mins)	Class Meeting (125 mins)	Module/Section
		MODULE 6: INTRODUCTION TO EQUATIONS
37	23	6.3 The Basic Percent Equation (Obj. A, B)
38		6.3 The Basic Percent Equation (Obj. C)
	24	6.4 Percent Increase and Percent Decrease (Obj. A, B)
39		6.5 Mark Up and Discount (Obj. A, B)
40	25	6.6 Simple Interest (Obj. A)
		Test #3 (** to be announced)
41	26	* Summary and Review for Final Exam
42	27	Comprehensive **Departmental Final Exam** on all course material covered

* Review Sessions before each class test/exam will only be done if time permits.

** Test #3 is optional. This is not a departmental test or exam. The instructor will decide whether to include this test, the content of the test, and notify the students appropriately.

PLEASE NOTE:

∞ **NO FORMULA SHEETS WHATSOEVER** (i.e., none provided by the instructor, nor by the student her/himself) – **ARE ALLOWED TO BE USED BY STUDENTS DURING ANY IN-CLASS TESTS OR EXAMS IN MTH 086.**

∞ **NO CALCULATORS OF ANY KIND** (i.e., no scientific calculators, no graphing calculators, no cell phone calculators, etc.) **ARE PERMITTED TO BE USED ON ANY IN-CLASS TESTS OR EXAMS IN MTH 086**.